Strong Metal–Support
Interactions

ACS SYMPOSIUM SERIES **298**

Strong Metal–Support Interactions

R. T. K. Baker, EDITOR
Exxon Research and Engineering Company

S. J. Tauster, EDITOR
Exxon Research and Engineering Company

J. A. Dumesic, EDITOR
Department of Chemical Engineering
University of Wisconsin—Madison

Developed from a symposium sponsored by
the Divisions of Petroleum Chemistry, Inc.,
Industrial and Engineering Chemistry,
and
Colloid and Surface Chemistry
at the 189th Meeting
of the American Chemical Society,
Miami Beach, Florida,
April 28–May 3, 1985

American Chemical Society, Washington, DC 1986

Library of Congress Cataloging-in-Publication Data

Strong metal–support interactions.

Includes bibliographies and index.

1. Metal catalysts—Congresses. 2. Catalysis—
Congresses.

I. Baker, R. T. K., 1938–   . II. Tauster, S. J.,
1935–   . II. Dumesic, J. A., 1949–   . IV. American
Chemical Society. Division of Petroleum Chemistry.
V. American Chemical Society. Division of Industrial
and Engineering Chemistry. VI. American Chemical
Society. Division of Colloid and Surface Chemistry.

QD505.S77   1986   541.3'95   85–30708
ISBN 0–8412–0955–3

ACS Symposium Series

M. Joan Comstock, *Series Editor*

Advisory Board

FOREWORD

The ACS SYMPOSIUM SERIES was founded in 1974 to provide a medium for publishing symposia quickly in book form. The format of the Series parallels that of the continuing ADVANCES IN CHEMISTRY SERIES except that, in order to save time, the papers are not typeset but are reproduced as they are submitted by the authors in camera-ready form. Papers are reviewed under the supervision of the Editors with the assistance of the Series Advisory Board and are selected to maintain the integrity of the symposia; however, verbatim reproductions of previously published papers are not accepted. Both reviews and reports of research are acceptable, because symposia may embrace both types of presentation.

CONTENTS

INDEXES

PREFACE

THE DISCOVERY OF STRONG METAL–SUPPORT INTERACTIONS in 1978 led to a renaissance in the study of the effects of supports in supported metal catalysts. The importance of the support has been appreciated for many years. In fact, the concept of electronic interactions between metal particles and semiconducting supports was the basis for intense research on supported metal catalysts in the 1950s. In the late 1960s and early 1970s, research on the effects of metal particle size on refractory supports such as SiO_2, Al_2O_3, and MgO became an active discipline. When it was found in 1978 that the properties of Group VIII metals could be altered dramatically by supporting these metals on reducible supports such as TiO_2, research on metal–support interactions became an active area once again. This book focuses on the advances that have been made recently in this area.

The origin of strong metal–support interactions has been the subject of much research during the past seven years. It has been proposed, for example, that these interactions are due to electronic interactions between the metal and the support, changes in a metal surface structure or particle size induced by the support, or the presence of support species on the metal surface. In general, any of these effects may be dominant for a given catalyst system as documented by the papers in this book. Undoubtedly, the origin of strong metal–support interactions for large metal particles is the presence of support species on the metal surface. This phenomenon is developed at length in this book by a number of papers that address the evidence for the presence of oxide species on metal surfaces and that probe the effects of these species on the catalytic and chemisorptive properties of the metal.

We hope that this collection of papers on metal–support interactions will serve as a "catalytic support" for further research and lead to new applications or to discoveries of new classes of metal–support interactions.

R. T. K. BAKER
Corporate Research Science Laboratories
Exxon Research and Engineering Company
Annandale, NJ 08801
S. J. TAUSTER
Corporate Research Advancement and Transfer
Exxon Research and Engineering Company
Annandale, NJ 08801
J. A. DUMESIC
Department of Chemical Engineering
University of Wisconsin—Madison
Madison, WI 53706

Strong Metal–Support Interactions
Facts and Uncertainties

S. J. Tauster

Corporate Research Advancement and Transfer, Exxon Research and Engineering Company, Annandale, NJ 08801

Studies of metal-support interactions have yielded significant progress in the past few years. It is appropriate, at the outset of the Metal-Support Interaction symposium, to review our current state of knowledge. We will focus on those facts that seem firmly established as well as on the major items that remain controversial.

A brief synopsis of the older literature will be given first. The finding that the chemisorption properties of Group VIII metals are drastically altered by titania supports following high-temperature-reduction (1), later generalized to include other reducible transition metal oxides (2), was interpreted to mean that a strong interaction occurred between these phases. A theoretical model (3) indicated covalent bonding between a Ti^{3+} cation and a Pt atom and further suggested that cation-to-metal charge transfer would strengthen the interaction. This electronic perturbation of the metal atoms was held responsible for the suppressed chemisorption of H_2 and CO, although the observation that even large metal particles suffered this suppression was difficult to explain on this basis. Evidence for the strength of metal-titania interactions came from electron microscopy which revealed that some metals (Pt (4), Ag (5)) had a tendency to spread on the reduced-titania surface. These interactions drastically suppressed activity for alkane hydrogenolysis (6,7); however, the same systems showed increased activity for $CO-H_2$ synthesis, often accompanied by improved selectivity to higher hydrocarbons.

TiO_x Overlayer Formation in Metal/Titania Model Systems

More recent research has greatly improved our understanding of these systems and has led to new questions as well. Of major importance is the discovery that reduced-titania (usually referred to as TiO_x) is surprisingly mobile and can migrate onto the surface of metals at temperatures commonly used for catalytic reaction or pretreatment. This remarkable structural dynamism introduces a new dimension to the interpretation of metal/titania systems.

0097–6156/86/0298–0001$06.00/0

As an example of this behavior, a 100Å-thick layer of Ni atop a layer of TiO_2 shows, initially, only Ni Auger peaks. Reduction at 723K of TiO_2 to TiO_x leads to the detection of both Ti and O signals. Evidently TiO_x has migrated a distance of 100Å, presumably along grain boundaries or the lateral surface of the Ni layer, appearing on top of the Ni ($\underline{8}$). If a layer of SiO_2 is interposed between the Ni and TiO_2, no Auger peaks of Ti or O are seen, indicating the absence of artifacts. Similar results have been obtained with a 30Å-thick Rh layer (Ti and O peaks seen with Auger spectroscopy and SIMS) ($\underline{9}$) and in another study involving Ni (120Å thick) on titania (Ti and O peaks seen with Auger spectroscopy) ($\underline{10}$). In the latter two cases, Ar ion sputtering led to a decrease in the Ti and O Auger signal intensities, indicating removal of the TiO_x overlayer.

The detailed mechanism of TiO_x-overlayer formation, occurring within minutes at moderate temperatures ($\underline{9}$) is still obscure. It must be understood, however, as a chemically-specific effect, driven by the interaction between the TiO_x and metal surfaces. The interaction is strong enough to compensate for the partial loss of Madelung energy caused by the disruption of the titania lattice. TiO_x migration is therefore not to be confused with ordinary sintering which leads to a decrease in the total surface area. Early experiments had in fact shown that high-temperature-reduction of metal/titania systems did not cause such a decrease (1).

The formation of TiO_x overlayers thus serves to demonstrate the existence of a metal-support interaction. The latter should be thought of, primarily, in materials-science terms; that is, a bonding between the phases. The effect of this interaction on catalytic properties is a separate question. Additional evidence for metal-titania interaction comes from three surface science studies in which metals were vapor deposited onto titania or $SrTiO_3$ surfaces pretreated so as to contain Ti^{3+} as shown by XPS or EELS. The important result was that surface Ti^{3+} was eliminated ($\underline{11},\underline{12}$) or decreased ($\underline{13}$) by the metal flux. These experiments, involving Pt, Ni or Pd as the deposited metal, point clearly to a metal-Ti^{3+} interaction. This finding appears to have been obscured by the inconclusive results of experiments aimed at establishing the degree of charge transfer in metal-titania interactions, usually involving XPS. Regardless of how the question of charge transfer is ultimately resolved, the fact of metal-Ti^{3+} interaction seems established by the results just cited.

THE EFFECT OF TiO_x-OVERLAYERS ON ADSORPTION-DESORPTION PROPERTIES

The significance of TiO_x-overlayer formation extends beyond the question of bonding between metals and titania. It suggests that the suppression of H_2 and CO chemisorption in metal/titania systems can be explained as simple site blockage due to the overlayer. Despite the appealing simplicity of this model the question is still open and evidence can be cited that points to a more complicated situation. Experiments have been performed under

conditions chosen to minimize the likelihood of overlayer formation. In one case a monolayer of Pt was vapor deposited at 130K on
a prereduced titania film. H_2 adsorption at 130K was determined
before and after annealing at 370K. The treatment reduced the
adsorption capacity by 2/3 although this low temperature would not
be expected to lead to an overlayer and, in fact, none was indicated by Auger spectroscopy(14). A similar result was obtained by
depositing Pt (0.5 ML) at 298K on a titania film that was either
oxidized or, alternatively, prereduced before deposition. Prereduction led to a 2/3 decrease in the amount of H_2 adsorbed at 120K
(15).

Another approach to the chemisorption-overlayer question
is to determine the degree of coverage with Auger spectroscsopy and
compare this with the amount of chemisorption suppression. Two
such studies have been reported and have reached opposite conclusions. In one case, varying amounts of Ti were deposited onto Pt
and then oxidized to TiO_x. H_2 and CO adsorptions were found to
vary linearly with TiO_x coverage and complete coverage was required
for complete suppression. Thus no effect other than simple site
blockage could be discerned (16). In another study Ni (120Å layer)
was deposited onto TiO_2. Reduction at 698K produced a TiO_x overlayer. The coverage of Ni, as well as its effect on CO adsorption,
was monitored as a function of time. At low TiO_x coverge between 5
and 9 Ni sites were deactivated for CO adsorption per Ti atom in
the overlayer (10,17).

If a TiO_x overlayer is able to do more than block sites,
i. e., if neighboring metal atoms are affected, this perturbation
should lead to changes in the Temperature Programmed Desorption
(TPD) spectra. Some studies have found this. For example, annealing a Pt/TiO_x sample at 370K lowered the H_2 desorption peak temperature 33K viz. annealing at 130K (14). Similarly, depositing Pt
(at 298K) on prereduced titania lowered the H_2 peak temperature 75K
viz. deposition onto an oxidized (TiO_2) film (15). Changes in the
CO TPD spectrum were seen as well. In both these studies TPD
changes were accompanied by other evidence of electronic perturbation of the Pt, i. e., suppressed adsorption in the apparent absence of TiO_x- overlayer formation (cited above). Correspondingly,
an experiment in which suppressed H_2 and CO adsorptions were attributed to simple site blockage of Pt by TiO_x (cited above) also
failed to detect any TPD changes induced by TiO_x (16). Finally,
the effect of a TiO_x overlayer on Ni was investigated (18). The
activation energy for CO desorption was found to be significantly
decreased, whereas in the case of H_2 both stronger and weaker binding states were induced by TiO_x.

PROPERTIES OF HIGH-TEMPERATURE-REDUCED (HTR) METAL/TITANIA CATALYSTS

TiO_x-overlayer formation has been amply demonstrated in
model systems amenable to surface science characterization. In the
case of well dispersed metal particles on high-surface-area titania,

firm evidence for overlayer formation is not available; still, one
might reasonably expect it to occur. As pointed out above, one
might attribute the near-total suppression of H_2 and CO chemisorption to near-total encapsulation of the metal.

Some recent reports, however, argue against such a dras-
tic interpretation. Although some degree of overlayer formation
may well occur following high temperature reduction of a metal/
titania catalyst, total coverage of the metal surface is not indicated. One example is the ability of the catalyst to generate
spillover hydrogen. Metal/titania catalysts that are reduced and
outgassed at 773K and subsequently contacted with H_2 at 298K show
rapid formation of hydroxyl groups (19) and Ti^{3+} (20) and an increase in electrical conductivity (21), resulting from the spill-
over-reduction of titania. In another study, the desorption isotherm of HTR Pt/titania, constructed by stepwise decreases in the
equilibrium pressure, extrapolated to H/Pt = 0.27 although the
conventional adsorption isotherm indicated H/Pt=0. The former
value agreed with the particle size derived from electron microscopy. Apparently, spillover hydrogen led to hydroxyl formation,
most of which occurred in the vicinity of the metal particles.

Spillover hydrogen can also be demonstrated through its
participation in catalytic processes. A recent report (22) described HTR Pt/titania as more active for acetone hydrogenation than
Pt/SiO$_2$. This reaction is believed to occur via coordination of
acetone to the oxide surface followed by attack of spillover
hydrogen.

In addition to supplying spillover-hydrogen, HTR metal/
titania catalysts have been reported to be active for dehydrogenation. (CO-H$_2$ synthesis reactions are not considered in this section due to the special problems they present). Despite the dras-
tic loss in ethane hydrogenolysis activity (1000-fold) caused by
high temperature reduction of Rh/titania the rate of cyclohexane
dehydrogenation decreased only ~ 20% (23) viz. the low temperature
reduced catalyst. It has also been reported that HTR Ni/titania is
about 0.5% as active for ethane hydrogenolysis as Ni/alumina but ~
50% as active for cyclohexane dehydrogenation (24). Other studies
(25) have yielded contrasting results, i. e., a sharp decrease in
dehydrogenation activity following high temperture reduction of
metal/titania. One may suspect that the details of catalyst pretreatment are important. It appears, however, that at least in
some cases activity can be severely suppressed for one reaction
while relatively unaffected for another, which runs counter to the
model of gross encapsulation of the metal particles.

For a given degree of TiO$_x$-overlayer coverge, it is
important to know how this phase is distributed on the metal surface. One possibility is "gross coverage", i. e., complete blockage of large areas of the metal with, perhaps, large metal areas
remaining unblocked. If the TiO$_x$ does not electronically perturb
nearby, unblocked sites, the unblocked portion of the metal would

be expected to exhibit normal adsorptive and catalytic properties. Thus a HTR 5% metal/titania catalyst with 80% gross coverage
of the metal by TiO_x should behave similarly to a 1% loading of the
same metal on, say, SiO_2. The above-cited changes in hydrogenoly-
sis-versus-dehydrogenation selectivity are not accounted for by
such a model. Further evidence against this simple picture comes
from the observation that Pt/titania and Pd/titania exhibit an
unusual relationship between H_2 and CO adsorptions, with the former
able to partially displace the latter (26,27). In addition, the
technique of "frequency response chemisorption" has been used to
demonstrate the existence of labile H_2 chemisorption sites in
Ni/titania and Rh/titania that are not found with the silica-supported metals (28).

Although gross coverage cannot account for the selective
suppression of hydrogenolysis viz. dehydrogenation activity cited
above, the possibility of a finely distributed TiO_x overlayer must
be considered. It has been pointed out (29) that this behavior is
characteristic of bimetallic systems (e.g., Cu-Ni) for which the
selective hydrogenolysis-suppressing effect of Cu is explained on
the basis of the large Ni ensembles required for this reaction.
Additional support for the metal/titania-bimetallic cluster analogy
comes from the faster disappearnce of bridged CO viz. linear CO
with increasing reduction of metal/titania (10,29,30). Also, H_2
reaction orders for hydrogenolysis are changed in the same way in
both types of systems (31). These observations point out the need
for information concerning the distribution, on an atomic scale, of
TiO_x moieties on the metal surface.

INFLUENCE OF THE METAL/TiO_x INTERACTION ON CO-H_2 SYNTHESIS PROPERTIES

Studies have shown that metal/titania catalysts often
show enhanced activity and/or selectivity for the CO-H_2 synthesis
reaction. This literature will not be reviewed here. Instead we
will concentrate on the question of how these reaction features
relate to the properties of metal/titania catalysts previously
discussed in this review. An important problem is that H_2O is a
by-product of the CO-H_2 reaction, raising the possibility that
reduced titania, "TiO_x", cannot exist in these systems. Since
there is much evidence that TiO_x is intrinsically related to the
results obtained in H_2O-free systems, the relationship between the
effects observed in these different environments is called into
question.

It must be remembered that, despite the presence of H_2O,
CO-H_2 synthesis reactions occur under net reducing conditions.
There will be a dynamic equilibrium between Ti^{3+} and Ti^{4+} governed
by the relative activities of all oxidants and reductants: H_2O,
CO_2, spillover hydrogen, CO and hydrocarbons. Indeed there is some
evidence that the properties associated with anhydrous metal/
titania systems can occur under H_2O-containing (but net reducing)
conditions as well. Suppressed CO chemisorption has been found

with in situ IR during $CO-H_2$ synthesis at 548K over Pt/titania (26) and Pd/titania (27). Suppressed $Ni(CO)_4$ formation has been found during $CO-H_2$ synthesis over NiFe/titania (32). The spreading of Fe on titania has been observed with TEM upon reduction at 773K in H_2 containing 1% H_2O (33). In another study of Fe/titania under simi-lar conditions, controlled atmosphere electron microscopy revealed that Fe atoms migrate from the center of the metal particle to re-duced titania created at the metal periphery by hydrogen spillover, resulting in a doughnut-shaped configuration (34). As a final note on this topic, a cyclic voltammetry study of $Pt/TiO_x/Ti$ showed suppression of CO chemisorption from a CO-saturated aqueous solu-tion (35).

The $CO-H_2$ synthesis properties of metal/titania catalysts have been found in several studies to be essentially unaffected by the temperature of reduction, in contrast to the strong effect this factor has on chemisorption properties. This problem has focused attention on the special nature of the metal-titania contact peri-meter. Reduction of titania, undoubtedly through hydrogen spill-over, begins there. It is important to note that Ti^{3+} cations are produced by reduction temperatures as low as 473K, as shown by O_2 adsorption/H_2O decomposition measurements (36) or by temperature programmed reduction. In the latter study, the amount of Ti^{3+} produced at temperatures below 503K was equivalent to a Ti^{3+}/Pt atom ratio of 0.6 (37).

These reduced cations may be involved in the creation of special CO adlineation sites which have been proposed to exist at the metal/titania contact perimeter (38-40). The concept of CO adlineation, i.e., of a CO molecule simultaneously coordinated to a cation and a metal atom, obviously demands an intimate association between these centers and thus, in fact, presupposes a bonding interaction between the metal and reduced titania. This could conceivably affect $CO-H_2$ synthesis properties in other ways. One is by simply inhibiting sintering of the metal particles. Another possibility would be the creation of metal sites near the contact perimeter with altered adsorption properties for H_2 viz. CO. Such sites might be capable of supplying an incresed flux of dissociated hydrogen to the metal surface. It should be noted that the H_2-displacing-CO effect, referred to above in connection with Pt/titania and Pd/titania, was also observed following reduction of these catalysts at 473K, although the effect was increased by re-duction at 773K.

Thus, it is quite conceivable that the metal-titania interaction influences the $CO-H_2$ synthesis reaction through effects that are concentrated at the contact perimeter and that, since reduction begins here, high temperature reduction is not required for changes in $CO-H_2$ synthesis properties to be observed. On the other hand, major suppression of H_2 (or CO) chemisorption, or of hydrogenolysis activity, requires that nearly the whole metal particle be affected. Even if suppression does not require blockage of all sites, i.e., if this can come about by the disruption of ensembles or the electronic perturbation of sites near the contact perimeter, overlayer formation will probably be required except, perhaps, in the case of very thin metal crystallites. It has been reported that the hydrogenolysis activity of a Rh/titania catalyst decreased 25% by reduction at 513K for 2 hours (41). This undoubtedly reflects the deactivation of sites near the contact perimeter. Deactivation increases continuously with temperature/time and becomes nearly total only following high temperature reduction.

To conclude this section, there is a need for a better understanding of the unusual $CO-H_2$ synthesis properties of metal/titania catalysts and related systems such as metal/niobia. The primary question to be answered in this regard concerns the stability of reduced titania in the $CO-H_2$ system. The fact that several reducible oxides (titania, niobia, vanadia, MnO, etc.) have been found to impart unusual $CO-H_2$ synthesis properties to supported metals suggests that support-reducibility is an important factor that is not cancelled by the $CO-H_2$ reaction environment.

SCOPE OF METAL-SUPPORT INTERACTIONS

Although most of this review has concerned metal/titania, it has been mentioned that other easily reducible oxides have similar support properties. Earlier publications pointed out that they differed as a class from main-group oxides such as alumina, silica and magnesia as well as from refractory (to reduction) transition metal oxides such as zirconia and hafnia (2, 42).

It appears, however, that support surfaces are not always as refractory to reduction as chemical intuition would dictate. An important new finding is that lanthanum oxide undergoes reduction, in the presence of a supported metal, to "LaO_x", and the properties of Pd/lanthana are similar in several respects to those of metal/titania (43,44). Even with alumina supports, surface reduction has been found in some instances (45-47). The reason for this anomalous behavior is not fully understood, although sulfur has been found capable of promoting the reduction (46). A recent report has described suppressed H_2 chemisorption on Rh/zirconia (48) although this was not found in an earlier study of Ir/zirconia (2). One may suspect differences in surface reducibility between the supports used in the two cases.

Although this review has dealt with the interactions of metals with reduced oxide surfaces, metal-support interactions are certainly not limited to these. Evidence for metal-support interaction involving non-reduced surfaces exists even in the metal/titania system. Enhanced hydrogenolysis activities have been found for low-temperature-reduced Rh/titania (7) and Ru/titania (49). These effects presumably involve interaction with Ti^{4+} ions.

REFERENCES

1. S. J. Tauster, S. C. Fung and R. L. Garten, J. Am. Chem. Soc. 100 170 (1978).
2. S. J. Tauster and S. C. Fung, J. Catal. 55, 29 (1978).
3. J. A. Horsley, J. Am. Chem. Soc. 101, 2870 (1979).
4. R. T. K. Baker, E. B. Prestridge and R. L. Garten, J. Catal. 59, 293 (1979).
5. R. T. K. Baker, E. B. Prestridge and L. L. Murrell, J. Catal. 79, 348 (1983).
6. E. I. Ko and R. L. Garten, J. Catal. 68, 223 (1981).
7. D. E. Resasco and G. L. Haller, Stud. Surf Sci. Catal.11, 105 (1982).
8. A. J. Simoens, R. T. K. Baker, D. J. Dwyer, C. R. F. Lund and R. J. Madon, J. Catal. 86, 359 (1984).
9. D. N. Belton, Y.-M. Sun and J. M. White, J. Am. Chem. Soc. 106, 3059 (1984).
10. S. Takatani and Y.-W. Chung, J. Catal. 90, 75 (1984).
11. Y.-W. Chung and W. B. Weissbard, Phys. Rev. B 20, 3456 (1979).
12. C.-C. Kao, S.-C. Tsai, M. K. Bahl, Y.-W. Chung and W. J. Lo, Surf. Sci. 95, 1 (1980).
13. Z. Bastl and P. Mikusik, Czech. J. Phys. B 34, 989 (1984).
14. D. N. Belton, Y.-M. Sun and J. M. White, J. Phys. Chem. 88, 5172 (1984).
15. D. N. Belton, Y.-M. Sun and J. M. White, J. Phys. Chem.88, 1690 (1984).
16. C. S. Ko and R. J. Gorte, J. Catal. 90, 59 (1984).
17. S. Takatani and Y.-W. Chung, Applications Surf. Sci. 19, 341 (1984).
18. G. B. Raupp and J. A. Dumesic, J. Phys. Chem. 88, 660 (1984).
19. J. C. Conesa, G. Munuera, A. Munoz, V. Rives, J. Sanz and J. Soria, Stud. Surf. Sci. Catal. 17, 149 (1984).
20. J. C. Conesa, P. Malet, G. Munuera, J. Sanz and J. Soria, J. Phys. Chem. 88, 2986 (1984).
21. J.-M. Herrmann and P. Pichat, Geterogen. Katal., 389 (1983).
22. J. Cunningham and H. Al Sayyed, Nouv. J. Chim. 8, 469 (1984).
23. G. L. Haller, D. E. Resasco and A. J. Rouco, Disc. Faraday Soc. #72, 109 (1981).
24. S. Engels, B.-D. Banse, H. Lausch and M. Wilde, Z. Anorg. Allg. Chem. 512, 164 (1984).
25. P. Meriaudeau, H. Ellestad and C. Naccache, Proc. 7th Int. Cong. Catal., E2 (1980).
26. M. A. Vannice, C. C. Twu and S. H. Moon, J. Catal. 79, 70 (1983).
27. M. A. Vannice, S.-Y. Wang and S. H. Moon, J. Catal. 71, 152 (1981).
28. G. Marcelin, G. R. Lester, S. C. Chuang and J. G. Goodwin, Actas Simp. Iberoam. Catal., 9th 1,271 (1984).
29. G. L. Haller, V. E. Henrich, M. McMillan, D. E. Resasco, H. R. Sadeghi and S. Sakellson, Proc. 8th Int. Cong. Catal. V, 135 (1984).
30. K. Tanaka and J. M. White, J. Catal. 90, 75 (1984).

31. E. I. Ko S. Winston and C. Woo, J. Chem. Soc., Chem. Commun., 740 (1982).
32. X.-Z. Jiang, S. A. Stevenson and J. A. Dumesic, J. Catal. 91, 11 (1985).
33. B. J. Tatarchuk and J. A. Dumesic, J. Catal. 70, 308 (1981).
34. B. J. Tatarchuk, J. J. Chludzinski, R. D. Sherwood, J. A. Dumesic and R. T. K. Baker, J. Catal. 70, 433 (1981).
35. J. Koudelka and M. Augustynski, J. Chem. Soc., Chem. Commun., 855 (1983).
36. D. Duprez and A. Miloudi, Stud. Surf. Sci. Catal. 11, 179 (1982).
37. T. Huizinga, J. Van Grondelle and R. Prins, Applied Catal. 10, 199 (1984).
38. R. Burch and A. R. Flambard, Stud. Surf. Sci. Catal. 11, 193 (1982).
39. R. Burch and A. R. Flambard, J. Catal. 78, 389 (1982).
40. J. D. Bracey and R. Burch, J. Catal. 86, 384 (1984).
41. D. E. Resasco and G. L. Haller, Applied Catal. 8, 99 (1983).
42. S. J. Tauster, S. C. Fung, R. T. K. Baker and J. A. Horsley, Science 211, 1121 (1981).
43. T. H. Fleisch, R. F. Hicks and A. T. Bell, J. Catal. 87, 398 (1984).
44. R. F. Hicks, Q.-J. Yen and A. T. Bell, J. Catal. 89 498 (1984).
45. A. C. Faro, J. Chem. Res. (S), 110 (1983).
46. K. Kunimori, Y. Ikeda, M. Soma and T. Uchijima, J. Catal. 79, 185 (1983).
47. S. W. Weller and A. A. Montagna, J. Catal. 20, 394 (1971).
48. T. Beringhelli, A. Gervasini, F. Morazzoni D. Strumolo, S. Martinengo, L. Zanderighi, F. Pinna and G. Strukul, Proc. 8th Int. Cong. V, 63 (1984).
49. G. C. Bond and X. Yide, J. Chem. Soc., Chem. Commun., 1248 (1983).

RECEIVED November 5, 1985

2

Information on Metal–Support Interactions from Near Edge X-ray Absorption Spectroscopy and Multiple Scattering Calculations

J. A. Horsley[1] and F. W. Lytle[2]

[1] Corporate Research Science Laboratories, Exxon Research and Engineering Company, Annandale, NJ 08801
[2] The Boeing Company, Seattle, WA 98124

We have used the L edge resonances in the x-ray absorption spectrum to obtain information on changes in the electronic structure of supported catalysts, compared to the bulk metal. The areas of the L_2 and L_3 edge resonances in silica and alumina-supported Ir catalysts indicate that the catalyst metal particles have a lower d orbital occupancy than the bulk metal. An approximate quantitative measure of the d orbital occupancy has been obtained using a calibration for a series of compounds based on multiple-scattering Xα calculations. At 90K the L_2 and L_3 edge regions in a silica-supported Pt catalyst show increased absorption 5-10 eV above the edge, compared to the same region in the bulk metal. We attribute this increased absorption to a transition to an empty antibonding state arising from the bonding of the metal particles to the support anions. Multiple-scattering Xα calculations on a simple cluster model of the metal-support interface support this explanation.

X-ray absorption spectroscopy has become an important tool for the characterization of catalysts. The near edge region of the x-ray absorption spectrum, within 50 eV of the absorption edge, can provide information on the electronic structure of the excited atom and the local structure around that atom which is largely complementary to the information provided by EXAFS (Extended X-ray Absorption Fine Structure). In particular, the absorption edge resonance lines or "white lines", at the L edges of transition metals, give information on the d orbital occupancy of the excited atom.[1,2] Changes in the intensity of these absorption edge resonances can be used to follow changes in the d orbital occupancy in supported catalysts.[1,3] We present here an analysis of the L_2 and L_3 edge structure in supported platinum and iridium catalysts using the results of multiple scattering Xα (MS-Xα) calculations of the L edge region in model clusters. Our analysis suggests that the observed changes in the L edge structure for

0097–6156/86/0298–0010$06.00/0
© 1986 American Chemical Society

these catalysts can be related to changes in the metal-support interaction.

Analysis of the L Edge Resonance Lines in Ir Catalysts

The L_2 and L_3 edges of supported Ir catalysts have been obtained by Lytle et al.[1] Fig. 1 shows the L_3 edges for 1 wt% Ir/SiO_2 and 1 wt% Ir/Al_2O_3 catalysts under He, O_2 and CO, super-imposed on the L_3 edge of Ir metal for comparison. It can be seen that the resonance lines for the "bare" (under He) supported catalysts have a significantly greater intensity than the reson-ance line of the bulk metal. There is a further increase in intensity upon O_2 chemisorption, but no significant increase upon CO chemisorption.

The L_2 resonance arises from transitions from the $2p_{1/2}$ core level to empty d states of $d_{3/2}$ character. Using the expres[2]sions relating the absorption cross sections of the L_2 and L_3 resonances to the d band vacancies given by Mattheis and Dietz[4], it can be shown that to a good approximation the total d orbital vacancies should be proportional to the sum of the L_2 and L_3 resonance areas. In order to relate the observed differences in L edge resonance intensity between Ir catalysts and the bulk metal to changes in d orbital occupancy we require a calibration that relates Xα orbital occupancies to the L_2 and L_3 resonance areas in a series of compounds. We have recently obtained accurate values of the d orbital occupancy for a series of platinum compounds using multiple-scattering Xα molecular orbital calculations.[2] The molecular orbital calculations were carried out on clusters representing the nearest neighbor environment of the Pt atom and included the effect of the 2p core hole. The area of the reson-ance was obtained by deconvolution of the edge into a Lorentzian function representing the transition to the empty d states and an arctangent step function representing the transitions to the con-tinuum. The total resonance area was taken to be the sum of the areas of the Lorentzians at the L_2 and L_3 edges. A plot of the d vacancies obtained from the molecular orbital calculations on the platinum compounds is shown in Fig. 2. It should be noted that this plot differs from the one given in Ref. 2, where the L_3 edge resonance area was erroneously multiplied by a weighting factor of 0.5. It can be seen that the relationship is indeed linear and that the d vacancies obtained from a cluster calculation on the pure metal can be included along with the values for the com-pounds.

We now use the relationship between d vacancies and L edge resonance area shown in Fig. 2 to obtain quantitative infor-mation on the d orbital occupancy in the Ir catalysts. We have found using SCF-Xα atomic wave functions that the oscillator strength for the 2p→5d transition in the Ir atom is very close to that of the corresponding transition in the Pt atom, so the relationship in Fig. 2 should be applicable to Ir compounds also. The resonance lines for the supported catalysts were

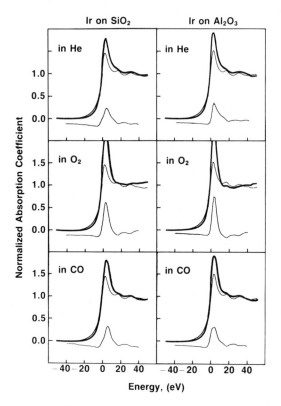

Fig. 1   L₃ absorption edges for iridium catalysts under various
conditions (heavier lines) aligned with the same edge in
the bulk metal. The difference spectrum shown below each
pair of edge spectra was obtained by subtracting the bulk
metal edge from the catalyst edge.

extracted from the edge structure by the same deconvolution as was used for the compounds. The fit obtained for the L_3 edge of the 1 wt% Ir/Al_2O_3 catalyst is shown in Fig. 3. The Lorentzian has a somewhat larger area in the catalyst than in the bulk metal, reflecting the more intense resonance line at the catalyst absorption edge. From the relationship in Fig. 2 the increase in area of the resonances at the L_2 and L_3 edges corresponds to an increase of 0.24 d vacancies over the bulk value in Ir/Al_2O_3 and an increase of 0.28 d vacancies in Ir/SiO_2.

The origin of the increase in d vacancies in the iridium catalysts is not clear. The average electronic configuration of an Ir atom in a small particle could differ significantly from that of an Ir atom in the bulk metal. However, we note that the white line at the L_3 edge of supported Pt catalysts does not show any significant increase in intensity compared to the bulk (although the line is broadened at low temperatures, which is discussed in more detail in the next section). If the increase in d orbital vacancies in the catalysts is attributed to the small particle nature of the Ir, then it is difficult to understand why a similar increase is not observed in small Pt particles of very similar average size. It does not seem probable, therefore, that the increase in d vacancies can be attributed to the small particle nature of the Ir in the catalysts, although further work is needed on the electronic structure of small Ir clusters in order to rule out this explanation with certainty.

Another possible explanation of the increase in d vacancies in the Ir catalysts is that d electrons are withdrawn from the metal particles by the support. However, it is difficult to understand how a large average charge transfer of 0.25 electrons per metal atom can be caused by interaction with wide band gap insulators such as SiO_2 and Al_2O_3. The valence band in these supports has few empty states that are capable of accepting the electrons transferred from the metal particles. A much more probable explanation is that some of the Ir cations survive the reduction treatment used to prepare the catalyst (the support was impregnated with chloroiridic acid and reduced for two hours at 773 K). The Ir cations would be anchored to the support through the oxygen atoms on the support surface to produce species of the type

It has been shown that Rh/Al_2O_3 catalysts prepared from $RhCl_3$ under similar conditions retained some chlorine, and IR spectroscopy of CO absorbed on these catalysts indicated that Rh cations were present.[5] ESR measurements have detected Rh^{2+} ions in reduced Rh/Al_2O_3 and Rh/TiO_2 catalysts.[6] Further work is required in order to determine whether oxidized species are also present in reduced Ir catalysts.

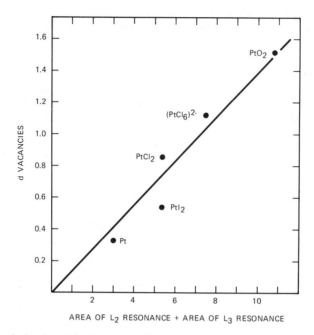

AREA OF L_2 RESONANCE + AREA OF L_3 RESONANCE

Fig. 2   Relationship between the unoccupied d states in platinum
metal and a series of platinum compounds, calculated in
the 2p core hole potential, and the sum of the areas of
the L_2 and L_3 resonance lines.

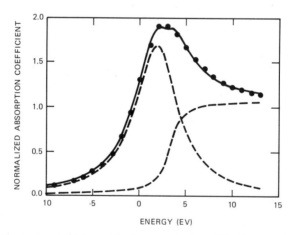

ENERGY (EV)

Fig. 3   Experimental L_3 edge for 1 wt % Ir/Al_2O_3 catalyst (filled
circles) together with the fit to the sum of the
Lorentzian and arctangent functions. The separate
components of this function are also shown.

Analysis of the L Edge Structure in Supported Pt Catalysts

We now consider the L edge region in supported Pt catalysts. Measurements of the L_2 and L_3 edge region in a 0.5 wt % Pt/SiO_2 catalyst have been carried out by Lytle et al.[7] at a series of different temperatures between 90 K and 773 K. Figure 4 shows the L_3 and L_2 edges in this catalyst under He and H_2 for the limiting temperatures of 90 K and 773 K. The corresponding edges for the bulk metal are shown for comparison. It can be seen that the L_3 edge white line in the catalyst at 90 K is much broader than the white line at the L_3 edge at the bulk metal because of a significant increase in the absorption cross section in the region 5-10 eV above the edge. However at 773 K the broadening has disappeared and the edge peak is now somewhat narrower than that in the bulk metal edge. Although the L_2 edge does not show a white line there is a corresponding increase in the absorption coefficient at 5-10 eV above the edge, in the catalyst at 90 K, for this edge as well. The change in the L_3 edge peak is continuous between the two temperatures and is completely reversible on cooling, showing that the change is not due to agglomeration of the Pt particles in the catalyst. The catalyst edge region at 90 K under H_2 is almost identical to the edge region under He, which means that the changes cannot be attributed to desorption of chemisorbed H_2. It is also highly improbable that changes in the region 5-10 eV above the edge, which corresponds to transitions to empty states that lie far above the Fermi level, can be attributed simply to the flow of electrons into the empty d band states in this region.

One possible explanation of the L edge changes in the Pt catalyst is that there is a change in the shape of the particles as the temperature is increased. The particles could have a flat raft-like shape at low temperature because of bonding to the support and assume a spherical or hemispherical shape at higher temperatures when the bonds to the support are broken. To test this explanation we have carried out calculations of the L edge absorption cross section for three different clusters of Pt atoms: a) a Pt_{13} cube octagon, representing the nearest neighbor environment in bulk Pt or a large three-dimensional particle, b) a Pt_{10} cluster representing the nearest neighbor environment of an atom on the close packed (111) surface of Pt, or in a two layer raft-like structure, and c) a Pt_7 cluster which represents the nearest neighbor environment of an atom in a two-dimensional single layer raft-like structure. The structures of the three clusters are shown in Fig. 5. The limitation of the clusters to the first shell of nearest neighbors might appear to be a fairly severe approximation. However, the short lifetime of the 2p core hole causes a substantial broadening (5 eV) of all the near edge features. Because of this broadening, features due to scattering from second and higher shells of nearest neighbors might not be resolved.

The cluster potentials were obtained by superimposing the self-consistent $X\alpha$ charge densities for the individual Pt

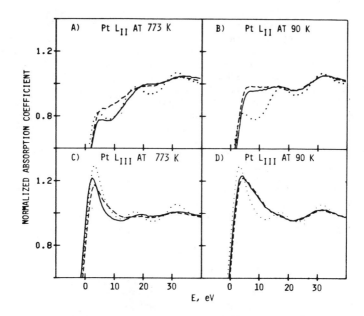

Fig. 4   L$_2$ and L$_3$ edge spectra for 0.5% Pt/SiO$_2$ catalyst under He
          (continuous line) and H$_2$ (dashed line), compared to the
          corresponding edge spectrum for the bulk metal (dotted
          line), at temperature of 773 K and 90 K.

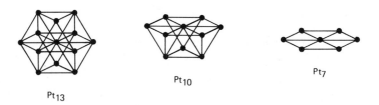

Fig. 5   Pt clusters for which adsorption cross section calculations
          were carried out.

atoms in the cluster. The L edge structure was obtained by calculating the absorption cross section for transitions from the 2p core level to continuum states above the ionization threshold of the cluster. The ionization threshold was made to coincide with the Fermi level of the cluster by surrounding the cluster with a negatively charged sphere of radius equal to the surrounding muffin tin sphere. In this way the empty d states below the vacuum level become continuum states and do not need to be treated separately. Absorption cross sections for the core to continuum transitions were obtained with the Ms-Xα method[8] using the program of Davenport,[9] modified to allow the use of a deep core level as the initial state. The spin-orbit splitting of the 2p core level was not included in the calculations, which must therefore be regarded as leading to an "average" of the L_2 and L_3 edge structure.

The calculated L edge structure for the central atom of the three Pt clusters is shown in Fig. 6. The calculated near edge structure for the Pt_{13} cluster reproduces fairly accurately the resonance line and the two weaker peaks in the bulk Pt near edge structure, indicating that the core hole broadening does indeed wash out features due to scattering from atoms outside the first shell. For the Pt_{10} and Pt_7 clusters the weak features above the resonance line are considerably diminished in amplitude and have almost disappeared completely in the case of the Pt_7 cluster, reflecting the weaker back scattering of the reduced number of nearest neighbors. However, the increased absorption in the region 5-10 eV above the edge, which leads to the broadening of the resonance line, is not obtained in any of the calculations. The calculations indicate that the changes in the L edge are not caused by a change of shape of the Pt particles.

The most probable explanation of the temperature dependence of the L edge structure in the catalyst appears to be that it arises directly from changes in the interaction of the metal clusters with the support anions. The interaction between the 2p orbitals of the support oxygen ions and the Pt 5d orbitals will produce bonding and antibonding hybrid orbitals below and above the Pt d band, respectively. The increased absorption at low temperature can be attributed to transitions to the empty antibonding states. At higher temperatures the interaction with the support is broken (which may be accompanied by a change in shape of the particles) and the edge peak becomes narrow, as would be expected from the calculations on the isolated clusters. Using the SCF-MS-Xα method we have calculated the orbital energy levels for a simple model of the metal-support interaction in the catalyst, namely a single Pt atom interacting with a $(SiO_4)^{8-}$ tetrahedral cluster. The geometry of the cluster, is shown in Fig. 7 together with the calculated orbital energies. The Pt-O distance used, 1.91 Å was the value obtained for the Pt-O distance in the catalyst from EXAFS.[7] There is a Pt5d-O2p antibonding level about 3 eV above levels that are largely non-bonding Pt 5d, (which will correspond to levels that are close to the Fermi level in a Pt cluster). An empty state at about 3 eV above the Fermi

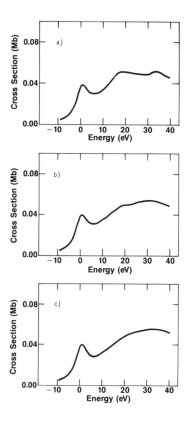

Fig. 6   Calculated near edge structure for the L edges of a)
         Pt$_{13}$, b) Pt$_{10}$ and c) Pt$_7$ clusters.

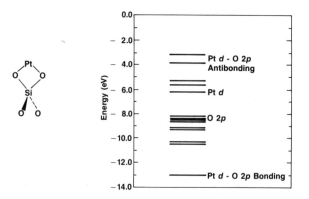

Fig. 7 Orbital energy levels for a (Pt SiO$_4$)$^{8-}$ cluster
        calculated using the SCF-MS-Xα method.

level of the Pt cluster would be in the right region to produce the extra absorption that broadens the edge peak. Although this simple cluster model only very approximately represents the real system the results of the calculation do provide some support for attributing the changes in the edge structure to changes in the metal-support interaction.

Finally, we note that the white line obtained by Short et al.[10] at the L_3 edge of a 1.7 wt % Pt/TiO_2 catalyst prepared by reduction at 450°C does not show any broadening. This catalyst showed suppression of H_2 chemisorption and was therefore in the SMSI (Strong Metal-Support Interaction) state. The metal-oxygen interaction in this catalyst appears to be much weaker than in Pt/SiO_2 following our explanation of the broadening of the white line, perhaps because it is dominated by interaction between the platinum clusters and the titanium ions.[11]

Conclusions

An analysis of the L edge resonance lines in supported Ir catalysts using Pt compound resonance lines as a calibration, indicates that the Ir particles in the catalyst have about 0.25 more d vacancies per atom than the bulk metal. We attribute this increase to the presence of cationic Ir species anchored to the support that have survived the reduction treatment. The corresponding edges in a 0.5 wt % Pt/SiO_2 catalyst show increased absorption (compared to Pt metal) in the region 5-10 eV above the edge, measured at 90 K, but this feature disappears at higher temperatures. Calculations of the absorption edge structure for several model Pt clusters indicate that the temperature dependence is not due to a change in shape of the Pt particles. Instead, we propose that it arises from changes in the interaction of the metal particles with the support anions.

References

1. F. W. Lytle, P. S. P. Wei, R. B. Greegor, G. H. Via and J. H. Sinfelt, J. Chem. Phys. 70 4849 (1979).

2. J. A. Horsley, J. Chem. Phys. 76 1451 (1982).

3. A. N. Mansour, J. W. Cook and D. E. Sayers, J. Phys. Chem. 88 2330 (1984).

4. L. F. Mattheis and R. E. Dietz, Phys. Rev. B 22 1663 (1980).

5. C. A. Rice, S. D. Worley, C. W. Curtis, J. A. Guin and A. R. Tarner, J. Chem. Phys. 74 6487 (1981).

6. T. Huizinga, Thesis, Technische Hogeschool, Eindhoven (1983).

7. F. W. Lytle, R. B. Greegor, E. C. Marques, V. A.
    Biebescheimer, D. R. Sandstrom, J. A. Horsley, G. H. Via and
    J. H. Sinfelt, in "The New Surface Science in Catalysis", ACS
    Symposium Series (to be published).

8. J. L. Dehmer and D. Dill, J. Chem. Phys, 65 5327 (1976).

9. J. W. Davenport, Phys. Rev. Lett. 36 945 (1976).

10. D. R. Short, A. N. Mansour, J. W. Cook, D. E. Sayers and J. R.
    Katzer, J. Catal. 82 299 (1983).

11. S. J. Tauster, S. C. Fung, R. T. K. Baker and J. A. Horsley,
    Science 211 1121 (1981).

RECEIVED October 30, 1985

Chemisorption and Catalysis over TiO_2-Modified Pt Surfaces

D. J. Dwyer, J. L. Robbins, S. D. Cameron, N. Dudash, and J. Hardenbergh

Corporate Research Science Laboratories, Exxon Research and Engineering Company, Annandale, NJ 08801

The migration of titania from the support onto a Pt surface during high temperature reduction is clearly demonstrated and correlated with chemisorption suppression. The origin of chemisorption suppression is a simple loss in surface adsorption sites; no electronic or chemical modification of the Pt occurs. The rate of methanation (3/1 H_2/CO, 1atm) over Pt blacks increases by well over 2 orders of magnitude when surface TiO_2 is added. This rate enhancement does not require high temperature reduction and is not directly related to chemisorption suppression. Ti^{+3} centers at the periphery of the titania-Pt interface are implicated in the mechanism of enhanced methanation.

The nature of strong metal support interactions (SMSI) has been the subject of debate and controversy over the past several years. SMSI is characterized by the loss of a supported metal's ability to chemisorb CO or H_2 following high temperature reduction.[1] In addition, the rate of methane synthesis from CO/H_2 can be enhanced over certain transition metals by using SMSI supports.[2-5] Pt/TiO_2 catalysts exhibit one of the largest SMSI effects in terms of enhanced methanation rates and have been extensively studied in the past.[6-12] Early explanations of the SMSI effect centered on an electronic model wherein the chemical properties of Pt were altered via a charge transfer between the support and the metal.[13,14] More recently, there is a growing body of evidence that a reduced form of titania migrates onto the metal surface during high temperature reduction.[15-20] In this work, we have studied SMSI effects in the Pt/TiO_2 system using a variety of surface science probes on both model and conventional catalysts. This work shows that TiO_2 in contact with Pt is readily reduced to a substoichiometric TiO_{2-x}. A substantial rearrangement of the electronic structure at the TiO_2/Pt

interface takes place during this reduction. However, this electronic reorganization does not induce chemisorption suppression via
a charge transfer mechanism as previously suggested. Rather, chemisorption suppression is due to the simple loss of sites associated
with the migration of TiO_{2-x} onto the Pt surface. When Pt blacks
are promoted by the addition of TiO_2 overlayers, a two order of
magnitude increase in the methanation rate is observed. The rate
enhancement does not require high temperature reduction and may be
associated with special interfacial reaction sites.

Experimental

The experimental apparatus used in this study is shown schematically
in Fig. 1. It consists of a multi-technique UHV surface science
chamber coupled directly to a reactor system. This configuration
permits surface characterization and reactor studies to be carried
out on both model and conventional catalyst systems. The reactor
and the down stream analytical train have been described elsewhere.(21,22)
The UHV surface probes used in this study were x-ray photoelectron spectroscopy (XPS), ion scattering spectroscopy (ISS) and temperature programmed desorption (TPD). XPS measurements were carried
out with Mg K_α radiation ($h\nu$ = 1253.6 eV) and a hemispherical electron energy analyzer set at 50 eV pass energy. ISS measurements
were made with the same analyzer but with reversed polarity on the
hemispheres and input lenses. The incident He_4^+ ion beam current
was 5 x 10^{-9} amps at 500 eV. This beam energy and ion current
caused no change in surface composition during the ISS measurements.
TPD experiments were carried out on a liquid nitrogen cooled sample
manipulator whose temperature could be linearly ramped at 10 Ks^{-1}
over the range 100 K to 1200 K. The desorbed molecules were mass
analyzed by a multiplexed quadrupole mass spectrometer.
Three types of samples were used in the study: Pt particles
supported on TiO_2 thin films, TiO_2 crystallites on Pt substrates and
0.8 wt % TiO_2 on Pt black. The TiO_2 thin films (200 nm film thickness) were prepared on a metal substrate heater assembly by chemical
vapor deposition (CVD). The films were characterized by x-ray diffraction and optical techniques. Pt was deposited on these thin
films in UHV via evaporation from a resistively heated Pt wire.
The TiO_2 overlayers on Pt substrates were prepared in the UHV
system; the details of this preparation are reported elsewhere.(12)
The TiO_2 modified Pt black was prepared by suspending two grams
of Pt powder (Johnson-Matthey Puratronic: 2 ppm Si, 1 ppm Fe) in a
solution of $Ti(O-C_9H_{19})_4$ (0.106g) in spectrograde heptane (4.0
ml). The heptane was evaporated in a nitrogen stream, the dry solid
heated in flowing air at 675 K for 2.5 hr and then in flowing 10%
H_2, 90% He at 450 K for 4 hr. The sample was passivated in air at
300 K and its surface area measured by Kr BET. X-ray diffraction
measurements showed no evidence for crystalline TiO_2 domains greater
than 50 Å.

Direct Evidence for the Migration of Titania

A model catalyst system idealized to permit meaningful TPD and ISS
studies of Pt/TiO_2 was prepared as shown schematically in Fig. 2.

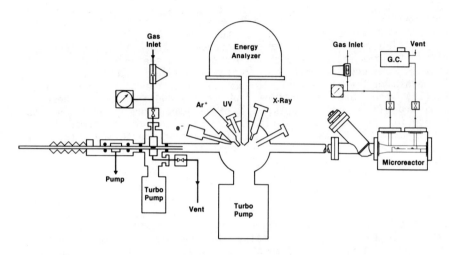

Figure 1. Schematic of the UHV surface analysis/mircoreactor system.

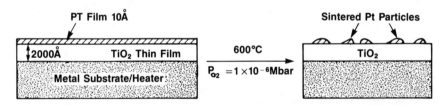

Figure 2. Schematic representation of the thin film model for the Pt/TiO$_2$ catalyst.

The TiO$_2$ thin films were first cleaned in UHV by Ar$^+$ ion sputtering followed by oxygen annealing (1100 K, 10^{-6} mbar O$_2$) to restore surface stoichiometry. A Pt film of approximately 10 Å (based on the attenuation of the Ti XPS signals) was vapor deposited on this defect free TiO$_2$ surface. The Pt film was then sintered by heating to 875 K (2 hr., 10^{-6} mbar O$_2$). The size of the Pt particles formed in this manner are estimated to be 2 to 5 nm in diameter based on comparison with microscopy studies in the literature.(18)

    CO TPD results from the presintered Pt/TiO$_2$ samples before and after high temperature reduction are shown in the left hand panel of Fig. 3. Desorption from the freshly deposited Pt particles (Fig. 3a) exhibits the characteristic two peak spectrum of polycrystalline Pt.(12) The lower temperature peak around 400 K is due to desorption from Pt (111) surfaces; the high temperature peak is due to desorption from steps and kinks.(23) Heating this surface in H$_2$ (10^{-6} mbar, 825 K) had a substantial impact on the CO TPD (Fig. 3b + 3c). There is a decrease in the total amount of CO chemisorption on the Pt particles and a change in the lineshape of the desorption spectrum. The decrease in chemisorption capacity is a direct manifestation of the SMSI effect. The changes in lineshape and peak maxima are probably associated with a change in particle morphology as suggested by White and coworkers(11) and observed by Baker et al.(24,25) The origin of chemisorption suppression in this model system can be directly ascertained by the highly surface specific technique of ISS. The ISS spectra associated with these TPD spectra are displayed in the right hand panel of Fig. 3. The spectrum labeled 3a is that obtained from the presintered Pt particles which exhibit normal chemisorption. A strong signal at a recoil energy of 0.88 is readily assigned by the binary collision model(26) to scattering from Pt surface atoms. The anticipated signals from Ti and O surface atoms are also clearly present in the spectrum. Upon exposure to high temperature hydrogen (Fig. 3b, 3c) there is a clear decrease in the Pt scattering signal concomitant with the decrease in CO chemisorption. The decrease in the ISS signal intensity indicates a loss of Pt scattering centers in the outermost atomic layer of the sample. It should be noted that the intensity of the XPS spectra (Ti, O, Pt) from the sample remained virtually constant during this sequence. The only possible explanation for these results is the creeping of titanium oxide onto the Pt surface during high temperature reduction.

    The SMSI induced chemisorption suppression could be reversed in these thin film samples by heating in oxygen. This fact is demonstrated in Fig. 4 where the left hand panel contains CO TPD results from the previously encapsulated samples after oxidation (875 K, P$_{O_2}$ = 10^{-6} mbar). It is clearly evident from the TPD results that the chemisorption capacity of the Pt particles could be restored by oxidation. In addition, the ISS results (right hand panel of Fig. 4) show the reemergence of Pt in the outermost atomic layer of the sample. In this manner it was possible to repeatedly cycle the sample into and out of the SMSI state. During the first reduction-oxidation cycle a permanent loss of some of the ISS and TPD signals occured. However, in all subsequent cycles both the ISS and TPD signal intensity could be fully recovered by oxidation. The initial permanent loss of some 10 – 15% of the chemisorption capacity is

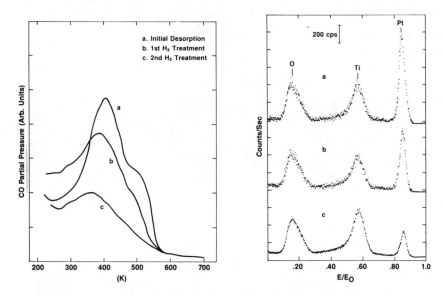

Figure 3. TPD and ISS results obtained from the thin film model catalyst as a function of reduction time at 875 K in 10^{-6} mbar H_2. a) 0 min. b) 15 min. c) 30 min.

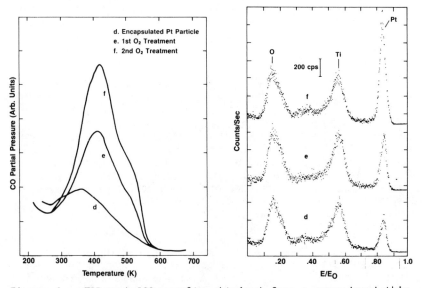

Figure 4. TPD and ISS results obtained from a prereduced thin film model as a function of oxidiation time at 875 K in 10^{-6} mbar O_2 d) 0 min. e) 20 min. f) 40 min.

probably due to the irreversible decoration of the Pt particles by
migrating titania. The subsequent reversible chemisorption suppression is due to titania that spreads across the Pt surface during
high temperature reduction and then spalls up on the surface during
oxidation.

The Nature of Chemisorption Suppression

The migration of titania onto the metal surface during high temperature reduction does not rule out chemical modification of the metal
as a factor in SMSI behavior. A second type of sample was studied
to probe the details of chemisorption suppression and to test
whether TiO_2 induces long range electronic effects on Pt. This
sample consisted of an ultra high purity Pt foil decorated with a
TiO_2 overlayer. The results of this study have been published in
detail elsewhere(12) but are included here for the sake of complete-
ness. This earlier study found that TiO_2 crystallites in contact
with Pt are readily reduced to a substoichiometric form by hydrogen
or carbon monoxide treatment in the vacuum chamber. The reduction
process results in a substantial rearrangement of the electronic
structure at the Pt/TiO_2 interface. These changes can be seen in
XPS spectra from these surfaces as shown in Fig. 5. The spectra
labeled 5a are those obtained from a freshly oxidized TiO_2 overlayer
on the Pt foil. The center panel contains the $2p^{1/2}$, $2p^{3/2}$ spin-
orbit doublet from titanium in the +4 oxidation state. Upon reduction with H_2 or CO a set of shoulders due to Ti^{+3} appear in this
spectral region (Fig. 5b). In addition to this true chemical shift,
there is a rigid shift ($\Delta E = 0.7 \pm 1$ eV) in the entire TiO_2 band
structure relative to the Pt fermi level. This rigid shift is
caused by a change in the fermi level position in the semiconducting
oxide and the establishment of an ohmic contact at the metal semiconductor junction.(12) Since the Ti^{+3} centers produced during the
reduction are donor states(27), this reorganization undoubtedly
involves charge transfer from the semiconductor to the metal. However, the charge transfer does not perturb the chemisorptive properties of the metal as was suggested in earlier studies.(13,14) This
fact can be seen in CO TPD results from these samples as a function
of TiO_{2-x} coverage. The left hand panel of Fig. 6 contains CO TPD
spectra at three coverages of TiO_{2-x}. The presence of TiO_{2-x} on the
surface only suppresses the total amount of chemisorbed CO. No new
chemisorption states are produced when TiO_{2-x} is added to the surface. The right hand panel of Fig. 6 demonstrates the linear relationship between the amount of chemisorption suppression and the
number of blocked surface sites as measured by ISS. It is clear
from these data that the presence of titanium oxide on Pt suppresses
chemisorption via a simple site blocking mechanism.

Methanation over TiO_2 Modified Pt

Chemisorption suppression via a simple site blocking mechanism is
not consistent with enhanced methanation rates observed for Pt/TiO_2
catalysts in the SMSI state.(2-5) A simple loss of CO adsorption
sites without a substantial change in the adsorption energy should
result in a decrease in catalytic activity. It is likely then that

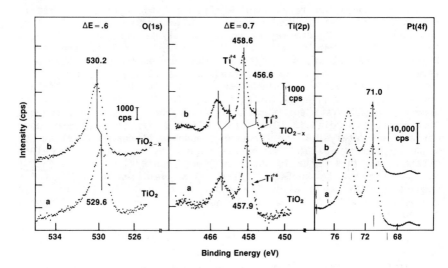

Figure 5. XPS data obtained from a TiO₂ on Pt polycrystal sample. a) Freshly oxidized sample b)After reduction by CO in the UHV chamber. Reproduced with permission from Ref. 12 (Fig. 3).

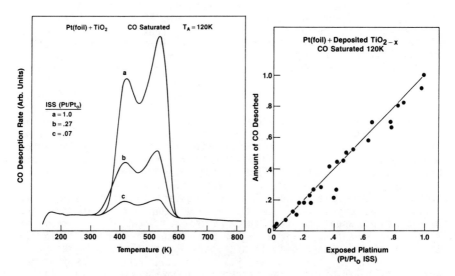

Figure 6. TPD of CO from Pt polycrystal as a function of TiO₂₋ₓ coverage. The fractional surface coverage of TiO₂₋ₓ was obtained by ratioing the Pt ISS signal to that of the clean surface Pt. Reproduced with permission from Ref. 12 (Figs. 8 and 10).

enhanced methanation rates over Pt/TiO$_2$ are not directly linked to
the SMSI effect as defined by chemisorption suppression. To study
the effects of TiO$_2$ overlayers on the catalytic response of Pt surfaces, Pt black samples with and without 0.8 wt. % TiO$_2$ were investigated. The 0.8 wt. % loading of TiO$_2$ was equivalent to 10 monolayers of TiO$_2$ if totally dispersed on the Pt black surface. The
methanation activities of these samples as a function of temperature
are shown in an Arrhenius format in Fig. 7. The reaction conditions
were 3:1 H$_2$:CO, latm and a maximum conversion of 3%. In agreement
with earlier studies,(8,10) clean Pt black was an extremely poor
methanation catalyst. The turnover frequency of our material at 548
K was approximately 10^{-5} molecules site^{-1} ·S^{-1} (assuming 8 x 10^{14}
site·cm^{-2}). This value is somewhat lower than earlier studies(8,10)
but was not due to surface impurities as verified by XPS and ISS
studies of the material. The addition of TiO$_2$ to the Pt black
causes a dramatic increase in the methanation activity (> 2 orders
of magnitude). It is important to note that the rates reported in
Fig. 7 are scaled by the physical surface area of the catalyst and
not by the Pt surface site density. Therefore, the rate increase
represents a true increase in the catalytic yield of the material.
This result is remarkable in light of the massive loading of surface
TiO$_2$ but is in excellent agreement with earlier studies of this
type.(10,28)

Surface analysis of the TiO$_2$ promoted Pt black revealed important information concerning the nature of the catalytic surface.
Fig. 8 contains the Ti 2p region of the XPS spectrum of the TiO$_2$
promoter after H$_2$ reduction (450 K, latm H$_2$). Two oxidation states
of titanium (+3 and +4) are readily visible in the spectrum. The
absolute binding energies are higher than those in Fig. 5 due to
differences in the location of the fermi level of the oxide.(12)
The presence of substantial Ti^{+3} character in the XPS spectrum indicates that TiO$_2$ in contact with Pt is readily reduced to a substoichiometric form even under mild reducing conditions (450 K).
These conditions are well below those required to induce titania
migration and chemisorption suppression in thin films. Another
important observation was obtained from ISS (see table insert in
Fig. 5), which shows that only 5% of the outermost atomic layers of
the sample is Pt. In spite of the extensive TiO$_{2-x}$ overlayer which
blocks 95% of the Pt adsorption sites, the methanation reaction rate
increased by over two orders of magnitude. The origin of this enhancement is not immediately evident from these measurements but
there is indirect evidence that Ti^{+3} centers located at the periphery of the titania-Pt interface may be important. This evidence
was obtained when the catalytic response of the TiO$_{2-x}$ promoted
sample was investigated at a higher temperature/higher conversion (~
5%) conditions. Prolonged (16 hr) operation of the TiO$_2$/Pt at 625 K
resulted in a 50% loss of catalytic activity. The deactivated material still exhibited linear arrhenius behavior with a slope indis-
tinguishable from that of the freshly charged catalyst (Fig. 9).
Surface analysis showed the activity loss was not due to carbon
deposition or the adsorption of any other impurity. In fact, ISS
and XPS revealed the deactivated catalyst contained less adventi-
tious surface carbon than the fresh one. Significantly, catalyst
deactivation accompanied both a 30% decline in the Ti^{+3} signal and a

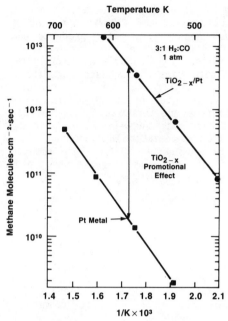

Figure 7. Methanation rate as a function of temperature obtained from Pt black compared to TiO_{2-x} modified Pt black. E_a 25 ± 1 kcal mole^{-1} for Pt, E_a = 20. ±1 kcal mole^{-1} for TiO_{2-x}/Pt.

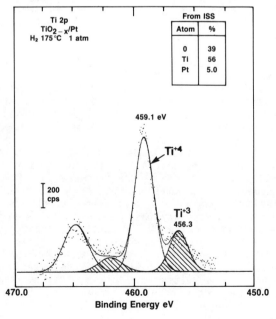

Figure 8. Ti XPS signal from TiO_{2-x}/Pt black sample after reduction.

doubling of the Pt ISS intensity (Fig. 10). These data clearly show
that methanation activity of the TiO_2/Pt does not correlate with Pt
metal surface area. Apparently, water produced under high conversion conditions is oxidizing the titania overlayer causing it to
spall up on the surface. The concomitant loss in catalytic activity
with increasing Pt site density suggests that the key chemistry is
occurring at sites associated with the Ti^{+3} centers and not on the
average Pt site. These sites presumably exist at the periphery of
the Pt titania interface and decrease in concentration as the oxide
overlayer contracts back into three dimensional crystallites. It is
important to note that even after 16 hr. under CO/H_2 at 5% conversion, Ti^{+3} is clearly evident by XPS. These results show that Ti^{+3}
centers and SMSI effects are quite durable under hydrocarbon synthesis conditions.

Conclusions

The data present in this paper support the encapsulation model (15)
for SMSI behavior. It is clear that reduced titania can migrate
onto the supported metal during high temperature reduction (see Fig.
3). This migration results in the suppression of chemisorption by a
simple physical blockage of surface sites. The chemical properties
of Pt adsorption sites which are not blocked remain unaltered by the
presence of reduced titania and exhibit normal chemisorptive behavior (Fig. 6). No evidence exists for electronic or chemical
modification of the Pt surface as is the case for alkali ad-
layers. (29-31) This simple loss of metal adsorption sites causes
chemisorption suppression and may account for reductions in rates of
ethane hydrogenolysis over SMSI catalysts. (32) On the other hand,
enhanced methanation rates over Pt/TiO_2 do not require high temperature reduction and are not directly linked to SMSI behavior (as
defined by chemisorption suppression). The higher methanation rates
appear to relate to some intrinsic property of the Pt/TiO_2 system
and not coupled to the migratory behavior of reduced TiO_{2-x}. The
origin of methanation rate enhancement still remains an enigma but
indirect evidence suggests that Ti^{+3} centers at the periphery of
metal particles are important. XPS data (Fig. 8) shows that TiO_2 in
contact with Pt is readily reduced to TiO_{2-x} even at mild reduction
temperatures (450 K). This reduction takes place well below the
temperature (> 650 K) where TiO_{2-x} migration is facile. The presence of Ti^{+3} centers at the periphery of the metal particles may
activate CO molecules via an unusual adsorption geometry as suggested by Vannice and others. (8,33,34) The proposed adsorption
geometry involves coordination of the oxygen end of Pt-CO to a Ti^{+3}
center. Presumably, this bonding weakens the CO bond and facili-
tates CO scission (methanation). However, no direct evidence for
this novel CO binding geometry was obtained in this study. No new
features were observed in the CO TPD spectrum of TiO_{2-x} modified Pt
(Fig. 6) and IR studies revealed no unusual CO stretching frequencies. (35) It could be that the concentration of such adsorption
sites is below our detection limits but the sites are extremely
active and dominate the catalysis. However this explanation is not
very satisfying and until more direct evidence is obtained, the
exact role of Ti^{+3} centers in CO hydrogenation remains speculative.

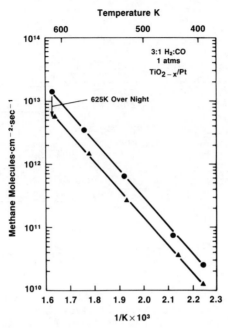

Figure 9. Arrhenius plot for TiO$_{2-x}$/Pt showing loss of activity at high temperature.

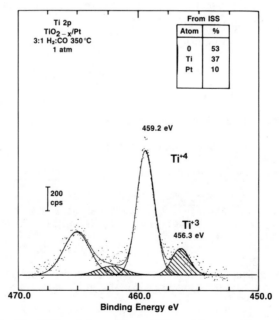

Figure 10. Ti XPS data from TiO$_{2-x}$/Pt sample after high temperature deactivation.

Literature Cited

1. Tauster, S. J.; Fung, S. C.; Garten, R. L. J. Am. Chem. Soc. 1978, 100, 170.
2. Vannice, M. A.; Garten, R. L. J. Catal. 1979, 56, 236.
3. Vannice, M. A.; Moon, S. H.; Twu, C. C. Prepr. Div. Pet. Chem. Amer. Chem. Soc. 1980, 25, 303.
4. Wang, S. Y.; Moon, S. H.; Vannice, M. A., J. Catal. 1981, 71, 167.
5. Vannice, M. A. J. Catal. 1982, 74, 199.
6. Vannice, M. A.; Twu, C. C.; Moon, S. H. J. Catal. 1983, 79, 70.
7. Vannice, M. A.; Twu, C. C. J. Catal. 1983, 82, 213.
8. Vannice, M. A.; Sudhakar, C. J. Phys. Chem. 1984, 88, 2429.
9. Ko, C. S.; Gorte, R. J. J. Catal. 1984, 90, 59.
10. Demmin, R. A.; Ko, C. S.; Gorte, R. J. J. Phys. Chem. 1985, 89, 1151.
11. Belton, D. N.; Sun, Y. M.; White, J. M. J. Phys. Chem. 1984, 88, 1690.
12. Dwyer, D. J.; Cameron, S. D.; Gland, J. Surface Sci., in press.
13. Horsley, J. A. J. Amer. Chem. Soc. 1979, 101, 2870.
14. Tauster, S. J.; Fung, S. C.; Baker, R. T. K.; Horsley, J. A. Science 1981, 211, 1121.
15. Meriaudeau, P.; Dutel, J. F.; Dufax, M.; Naccache, C., in "Studies in Surface Science and Catalysis", Vol 11, 1982.
16. Jiang, X-Z; Hayden, T. F.; Dumesic, J. A. J. Catal. 1983, 83, 168.
17. Belton, D. N.; Sun Y. M.; White, J. M. J. Phys. Chem. 1984, 88, 5172.
18. Simoens, A. J.; Baker, R. T. K.; Dwyer, D. J.; Lund, C. R. F.; Madon, R. J. J. Catal. 1984, 86, 359.
19. Chung, Y. W.; Xiong, G.; Kao, C. C. J. Catal. 1984, 85, 237.
20. Sadeghi, H. R.; Henrich, V. E. J. Catal. 1984, 87, 279.
21. Dwyer, D. J.; Hardenburgh, J. H. J. Catal. 1984, 87, 66.
22. Dwyer, D. J.; Hardenburgh, J. H. Appl. Surface Sci. 1984, 19, 14.
23. McClellan, M. R.; Gland, J. L.; McFeeley, F. R. Surface Sci. 1981, 112, 63.
24. Baker, R. T. K.; Prestridge, E. B.; Garten, R. L. J. Catal. 1979, 59, 390.
25. Tauster, S. J.; Fung, S. C.; Baker, R. T. K.; Horsley, J. A. Science 1981, 211, 1121.
26. Suurmeijer, E. P. Th. M.; Boers, A. L. Surface Sci. 1973, 43, 309.
27. Grant, F. A. Rev. Mod. Phys. 1959, 31, 646.
28. Kugler, E. L.; Garten, R. L., U. S. Patent # 4,273,724, 1981.
29. Garfunkel, E. L.; Crowell, J. E.; Somorjai, G. A. J. Phys. Chem. 1982, 86, 310.
30. Garfunkel, E. L.; Crowell, J. E.; Somarjai, G. A. Surface Sci. 1982, 121, 303.
31. Sette, F.; Stohr, J.; Kollin, E. B.; Dwyer, D. J.; Gland, J. L.; Robbins, J. L.; Johnson, A. L. Phys. Rev. Lett. 1985, 54, 935.
32. Resasco, D. E.; Haller, G. L. J. Catal., 1983, 82, 279.

33. Sachtler, W. M. H.; Shriver, D. F.; Hollenberg, W. B.; Lang, A. F. J. Catal. 1985, 92, 429.
34. Burch, R.; Flambard, A. R. J. Catal. 1982, 78, 389.
35. Robbins, J. L., unpublished data.

RECEIVED September 17, 1985

4

Effects of TiO$_2$ Surface Species on the Adsorption and Coadsorption of CO and H$_2$ on Ni

Implications for Methanation over TiO$_2$-Supported Ni

G. B. Raupp[1] and J. A. Dumesic[2]

[1]Department of Chemical and Bio Engineering, Arizona State University, Tempe, AZ 85287

[2]Department of Chemical Engineering, University of Wisconsin, Madison, WI 53706

Temperature-programmed desorption (TPD) under ultra-high vacuum conditions of CO and H$_2$ from polycrystalline nickel surfaces containing varying amounts of titania showed that the effects of titania are short-ranged. Carbon monoxide adsorption was weakened while hydrogen adsorption was strengthened with increasing concentration of surface titania. Coadsorption measurements confirmed that hydrogen and CO compete for sites on nickel. Using adsorption and desorption kinetic parameters derived from the TPD experiments, calculations indicated that shifts in heats of adsorption due to the presence of titania may result in a one to two order of magnitude higher hydrogen adatom coverage under steady-state methanation conditions relative to a clean Ni surface. This increased hydrogen coverage leads, in turn, to a one to two order of magnitude higher methanation turnover frequency.

We have recently probed the effects of titania surface species on the chemisorptive properties of a polycrystalline nickel foil using temperature-programmed desorption (TPD) of carbon monoxide and hydrogen under ultra-high vacuum (UHV) conditions (1). This work showed that submonolayer quantities of titania adspecies not only blocked adsorption sites, but also altered adsorption on neighboring nickel sites. Carbon monoxide adsorption strength was weakened, while hydrogen adsorption on nickel sites was only moderately perturbed. In addition, an activated state of adsorbed hydrogen was created and was thought to be associated with spillover of hydrogen from nickel to titania. These results were in agreement with the chemisorption properties of titania-supported nickel particles and thus strongly suggested that so-called strong metal-support interactions may be induced by the presence of titania adspecies on the metal surface.

Two issues were investigated in the present work. First, the relative importance of long-range (greater than first or second

nearest neighbor) electronic interaction versus short-range
geometric effects of the titania surface species on nickel were
studied. Second, the unresolved issue of how these titania
adspecies lead to enhanced CO hydrogenation activity for titania-
supported nickel catalysts relative to "clean" nickel crystallites
was addressed.

Results and Discussion

Carbon Monoxide Desorption. The details of CO and H_2 adsorption on
Ni as a function of titania coverage were examined through TPD to
differentiate between long and short-range interactions. In the
extremes, if long-range electronic interactions dominate,
incremental addition of titania adspecies will effect all adsorption
sites equally; however, if local interactions are important, new
binding states will be observed near titania while the remainder of
the surface sites will remain unperturbed.

The influence of different coverages of predeposited TiO_x on CO
thermal desorption from CO-saturated surfaces is shown in Figure 1.
For the clean Ni surface, CO adsorbs molecularly and desorbs in
first-order fashion with a peak maximum near 460 K, denoted as the
α_1(Ni) state, and a low-temperature shoulder at ca. 250-350 K,
denoted as the α_2(Ni) state. With increasing concentration of
surface titania species, four changes can be observed in the CO TPD
traces: (i) the population of the α_1(Ni) state is progressively
decreased, (ii) the population of desorption states near 300-400 K
is enhanced, (iii) a new desorption peak at ca. 200 K is created and
fills to progressively higher coverages, and (iv) the peak
temperature of the α_1(Ni) state is gradually shifted downward.
Experiments on an oxidized titanium foil showed that the low-
temperature peak at 200 K is due to CO adsorption on the titania
itself (2). No other states were observed on titania, and thus the
enhanced adsorption near 300-400 K range is attributed to adsorption
on Ni sites perturbed by nearby titania, denoted $\alpha_1(TiO_x/Ni)$.

The shift of the α_1(Ni) peak by about 20 K relative to the clean Ni
surface at a titania coverage of 0.1 monolayer may be taken as
evidence for an electronic interaction (3). However, this small
shift represents only a 6 kJ·mol^{-1} decrease in the activation energy
for desorption of CO from the value of 134 kJ·mol^{-1} on clean Ni. In
contrast, the desorption energy for the $\alpha_1(TiO_x/Ni)$ state can be
estimated to be 91 kJ·mol^{-1}. It is concluded, therefore, that long-
range electronic effects, although apparently present, are of
secondary importance to short-range effects.

Initial sticking coefficients (S_0) of CO at 150 K were estimated
from the initial slopes of CO coverage versus exposure plots for
each of the surfaces studied. Figure 2 shows the dependence of S_0
on titania coverage. From the initial slope of this curve we
estimate that each titania moiety affects approximately six Ni
adsorption sites, in agreement with the work of Takatani and Chung
(4). Allowing for error in absolute TiO_x coverage of ±40% and
assuming titania is well dispersed, i.e., does not form islands, it

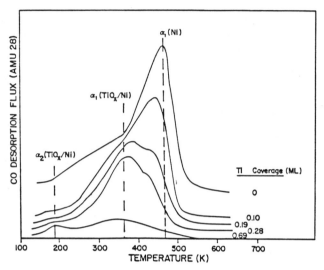

Figure 1. Effect of varying TiO_x precoverage on CO thermal desorption from Ni.

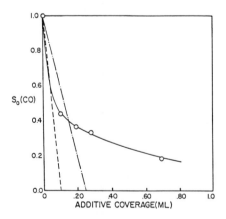

Figure 2. Variation of the initial sticking coefficient of CO, S_o, as a function of titania coverage. Long dash–dot line and short dashed line represent the theoretical proportionality of S_o to $(1-4\theta_{Ti})$ and $(1-10\theta_{Ti})$, respectively.

can be concluded that each TiO_x moiety affects not less than 4 and not more than 10 Ni surface sites (these limits are indicated in the figure). These limits are consistent with local modification of the Ni surface by the TiO_x moieties.

As demonstrated in recent HREELS and TPD studies of CO adsorbed on Ni (100), the majority of this local interaction may be explained by geometric effects or simple site blocking (5,6). A schematic diagram of a clean Ni(100) surface and a Ni(100) surface covered with a p(2x2) sulfur overlayer are shown in Figure 3. On clean Ni, CO adsorbs linearly on a single Ni atom, or atop site up to a coverage of θ_{CO} = 0.5 monolayer (6,7). Above this coverage so-called compression (8) or "domain" structures form (9,10), with further CO adsorption occurring at two-fold, or bridging sites. The presence of 0.25 monolayer S in a p(2x2) superstructure completely blocks all atop sites. Adsorption of CO can then occur only at bridging or four-fold hollow sites. The important conclusion is that the presence of S adatoms allows CO to adsorb at sites that would not otherwise be occupied due to CO-CO repulsive interactions.

Gland et al. (6) assigned characteristic CO desorption peak temperatures to each of the aforementioned sites, as reproduced in Figure 3. The similarity of the present CO TPD results with those of Gland et al. suggests that dispersed titania adspecies block CO adsorption on strongly-held, linear atop sites and allow adsorption onto less-strongly-bound hollow or bridging sites. Moreover, this interpretation is consistent with a recent HREELS study by Takatani and Chung (4) for CO adsorption on a nickel film deposited on titania. Following extended high temperature reduction, during which time the authors concluded that titania migrated onto the nickel surface, HREELS measurements showed no CO stretching bands near 2050 cm^{-1} (atop sites) nor near 1920 cm^{-1} (bridging sites). Instead, only a band at 1850 cm^{-1} was observed, which was assigned to "a site near the surface titanium oxide on nickel" (4). This stretching frequency is similar to the peak at 1815-1845 cm^{-1} observed on Ni(111) attributed to a threefold site (11) (note that there are no fourfold hollow sites on Ni(111)). The lack of precise structural characterization of the surface studied presently precludes unambiguous site assignment, of course. Nonetheless, the present results are consistent with the suggestion that the effects of surface titania with respect to CO adsorption are short-ranged and can largely be explained in geometric terms. Electronic effects, although present, are apparently of secondary importance. This means that for large supported Ni crystallites, TiO_x surface species must be present on the metal for these crystallites to exhibit the chemisorption behavior indicative of so-called strong metal-support interactions.

Hydrogen Desorption. Hydrogen desorption spectra as a function of increasing titania adspecies coverage on Ni for saturation hydrogen coverages obtained by exposing the samples to H_2 at 150 K are shown in Figure 4. Increasing amounts of titania decrease the population of the β_1 and β_2 dissociatively adsorbed states characteristic of clean Ni, with perhaps some preferential blocking of the β_2 state.

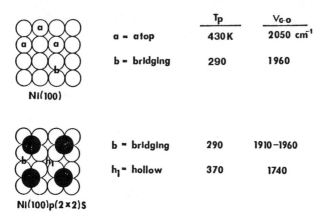

		Tp	V$_{G-O}$
	a = atop	430 K	2050 cm^{-1}
Ni(100)	b = bridging	290	1960
	b = bridging	290	1910–1960
Ni(100)p(2×2)S	h$_1$ = hollow	370	1740

Figure 3. Schematic of CO adsorption sites on clean and S-
covered Ni(100). Data from (6).

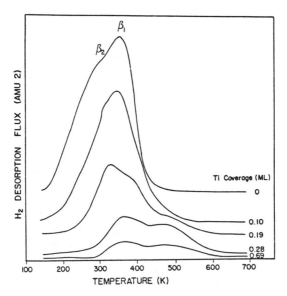

Figure 4. Effect of varying TiO$_x$ coverage on H$_2$ thermal
desorption from Ni.

A gradual upward shift in peak temperature for the two observed adsorption states with increasing titania coverage is evident. Because desorption of each observed state is described by recombination of hydrogen atoms, or second-order kinetics, a portion of the upward shift in peak temperature is due to lower initial adsorbate coverages. Kinetic analysis of the desorption traces revealed that overall hydrogen adsorption strength is increased by the presence of titania adspecies. Assuming a pre-exponential of 10^{-6} m^2·molecules^{-1}·s^{-1}, estimated desorption activation energies increased from 86 and 66 kJ·mol^{-1} for the β_1 and β_2 states, respectively, on the clean Ni surface, to 108 and 75 kJ·mol^{-1} for the two states on the Ni surface containing 0.69 monolayers titania.

If the H$_2$ dosing temperature was raised to 300-400 K, slow filling of a more-strongly-bound adsorption state occurred, as evidenced by a new peak with desorption temperatures near 500-550 K (desorption activation energy of ca. 120 kJ·mol^{-1}). This behavior is attributed to the presence of an activation energy barrier for adsorption into this strongest state; the barrier is thought to be associated with the spillover of H atoms from Ni onto the titania itself. The interpretation that H adatoms spillover and adsorb on titania is supported by the results represented in the top half of Figure 5. For high coverages of titania and large exposures of H$_2$ at 300 K, the total coverage of hydrogen, including the activated state, is within 20% of the coverage on the clean Ni surface.

These results for hydrogen adsorption are consistent with a short-range influence of titania on nickel. The lower half of Figure 5 shows that at low TiO$_x$ coverage the initial sticking coefficient (S$_0$) for hydrogen adsorption decreases, with S$_0$ proportional to between $(1-4\theta_{Ti})^2$ and $(1-10\theta_{Ti})^2$; note that the effect of titania adspecies on hydrogen sticking coefficient is greater than for CO since H$_2$ requires two adjacent Ni sites for dissociative adsorption.

In summary, the presence of titania surface species weakens CO adsorption strength on Ni but increases hydrogen adsorption strength on Ni. In addition, a more-strongly-bound, activated adsorption state is created which is probably associated with spillover onto titania. This behavior is in marked contrast to that exhibited by either electronegative adatoms (S, Cl, P) on Ni, which weaken both CO and hydrogen adsorption, or electropositive adatoms (K, Na, Cs) which increase CO adsorption strength. Clearly one cannot rationalize observed shifts in adsorption strength consistently in terms of electronic interactions. Instead, geometric and structural factors must be controlling. For CO adsorption, it appears titania blocks adsorption at strongly-binding, linear atop sites and forces adsorption to occur at more-weakly-binding, multifold sites. For hydrogen, we speculate that the observed increases in binding energy could be a consequence of direct bonding of hydrogen adatoms to titania at nickel/titania interfaces (e.g., bridge bonding to a nickel atom and to an oxygen anion or Ti^{3+} cation). Alternatively, the higher apparent binding energy could be due to steric hindrance to hydrogen adatom migration and recombination, prerequisite steps for molecular H$_2$ desorption.

Coadsorption of Hydrogen and CO Under UHV Conditions. Desorption
spectra for hydrogen and CO coadsorbed at 130 K on the clean
polycrystalline Ni surface are shown in Figure 6. The surface was
first exposed to 5 L of H_2 (L = Langmuir = 1.33×10^{-4} Pa·s) and then
to varying CO exposures. Exposure to background CO resulted in
negligible CO desorption and H_2 desorption from β_1 and β_2 states
(atomic states) characteristic of low-index crystal planes of clean
Ni. For CO exposures from 0.5 to 3 L, hydrogen adsorption is
weakened, a portion of the more-strongly-bound β_1-state being
transformed to the weaker β_2-state. In addition, a third desorption
state becomes evident as a shoulder on the low-temperature side of
the desorption trace. This corresponds to the appearance of a
similar peak in the CO desorption spectrum. These states are
strongly interacting, and were originally termed Σ-states by Goodman
et al. (12) in their work on Ni(100). Since the Σ-states are not
formed on Ni(111) (13,14), but are filled to a greater population on
Ni(100) than observed here, it appears that the polycrystalline Ni
foil consists of both (100) and (111) crystalline surface planes.

For the polycrystal exposed first to near-saturation doses of CO and
then to H_2, hydrogen adsorption into the β_2-state was completely
blocked. In addition, no Σ-states were formed. Consistent with
published spectroscopic characterization which showed that no C–H or
O–H bonds are formed under UHV coadsorption conditions, no CO
hydrogenation products could be detected desorbing from the surface
during heating.

These results favor a model originally proposed by Koel et al. (15)
wherein CO and H compete for the same adsorption site corresponding
to the most-strongly-bound states (α_1-CO and β_1-H). The α_1-CO state
has a higher heat of adsorption and hence displaces β_1-H states to
either β_2 or Σ sites upon adsorption. Saturation of the β_1-H states
completely blocks adsorption of CO into the weaker α_2-CO sites
(desorption temperatures from 300–400 K).

The presence of ca. 0.2 monolayers of titania markedly alters H_2/CO
coadsorption behavior on the nickel surface. Figure 7 shows TPD
spectra for hydrogen and carbon monoxide coadsorbed on the titania-
containing surface. Pre-exposure to 10 L H_2 at 130 K gave an
initial H coverage near 80% of the saturation value, with desorption
occurring from three states. The lower energy β_1(Ni) and β_2(Ni)
states can be attributed to adsorption at Ni sites. The highest
energy, or β_2(TiO$_x$/Ni) state, is assigned to adsorption at a Ni site
perturbed by nearby titania, possibly interacting with titania at
nickel/titania interfaces. No alteration of the H_2 desorption trace
was evident upon exposure to CO corresponding to less than about 50%
of saturation CO coverage. For increasing CO coverages above this
level, β_1(Ni)-H states were displaced to β_2(Ni)-H sites. The most-
strongly-bound hydrogen state remained essentially unperturbed by
the presence of coadsorbed CO.

If the surface was first preexposed with CO to saturate the α_1(Ni)
state, subsequent adsorption of hydrogen was markedly supressed.
The β_1(Ni) state was nearly completely blocked and the β_2(Ni)

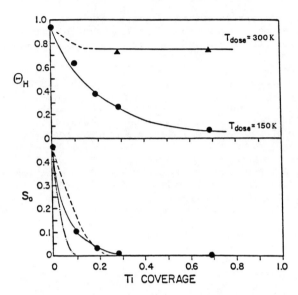

Figure 5. Variation of total H coverage, Θ_H, for hydrogen exposure at 150 and 300 K and initial sticking coefficient of H_2 at 150 K as a function of titania coverage. Short dashed line and long dash-dot line represent the theoretical proportionality of S_o to $(1-4\theta_{Ti})^2$ and $(1-10\theta_{Ti})^2$, respectively.

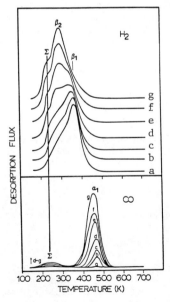

Figure 6. Desorption flux of H_2 and CO coadsorbed on a clean nickel surface. CO exposures were (a) background, (b) 0.5, (c) 0.8, (d) 1.5, (e) 3.0, (f) 5.0, and (g) 10 L.

state was partially blocked and slightly weakened. Fractional filling of the most-strongly-bound hydrogen state occured despite the presence of strongly-bound CO, and displaced ca. 20% of the α_1 (Ni) states to more-weakly-bound states.

Under no conditions were Σ-states observed in the 200-300 K regions of the desorption spectra. It is interesting to note, however, that the weaker α_2(Ni)-CO and β_2(Ni)-H states fill to greater populations than can be achieved when CO or H_2 are exposed alone, and that these states desorb at approximately the same peak temperature.

The fact that Σ-states are not observed on the titania-containing surface does not necessarily suggest a weaker interaction between H-adatoms and adsorbed CO at reaction conditions. Indeed, the catalytic importance of the low-temperature Σ-states has not been demonstrated, since there is no detectable difference in the methanation rate for Ni(111), Ni(100), and polycrystalline Ni (16,17), surfaces which exhibit markedly different coadsorption behavior with respect to the Σ-states.

The TPD data discussed above establish that H_2 and CO compete for adsorption sites on a polycrystalline nickel surface. Importantly, the presence of titania adspecies leads to more competitive hydrogen adsorption in the presence of coadsorbed CO. This behavior had been suggested earlier by Vannice (18).

Coadsorption Under Moderate Pressure Conditions. A Langmuir adsorption model was employed to estimate steady-state H-adatom and CO surface concentrations at moderate pressure and temperature conditions in the absence of reaction between these adsorbed species. Adsorption and desorption rates of CO and H_2 were equilibrated according to the following equations:

$$\emptyset_{CO}(P,T)\ S^o_{CO}(1 - \Theta_{CO} - \Theta_H) = \Theta_{CO} L\ \nu_{CO}\ \exp(-E_{CO}/RT)$$

$$2\emptyset_{H_2}(P,T)\ S^o_{H_2}(1 - \Theta_{CO} - \Theta_H)^2 = \Theta_H^2\ L^2\ \nu_{H_2}\ \exp(-E_{H_2}/RT)$$

where Θ_i = fractional coverage of surface by adsorbate i, S^o_i = initial sticking coefficient of adsorbate i, $\emptyset_i(P,T)$ = collision frequency of i, ν_i = preexponential factor for desorption of i, E_i = desorption activation energy of i, and L = Ni surface site density. Adsorption and desorption kinetic parameters were obtained from UHV TPD experiments.

Figure 8 compares the calculated CO and H coverages on Ni and TiO_x/Ni as a function of temperature at a total pressure of 760 Torr and an H_2/CO ratio of 4. The figure reveals that on clean Ni, the surface is essentially saturated with CO for temperatures less than 600 K. At higher surface temperatures, the CO coverage decreases as molecular CO desorption becomes important. Titania adspecies weaken CO adsorption and allow significantly higher H-adatom surface concentrations (1-2 orders of magnitude for temperature less than 650 K).

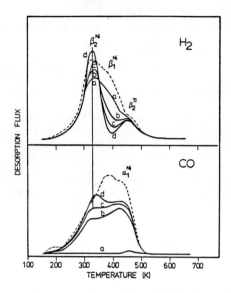

Figure 7.   Desorption flux of H_2 and CO coadsorbed on a titania-containing Ni surface.   Solid curves are for CO exposures of (a) background, (b) 7 L, (c) 10 L (d) 20 L.   Dashed lines correspond to desorption of a saturation layer of each gas adsorbed separately.

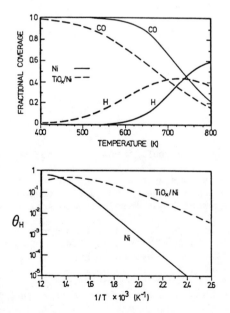

Figure 8.   Comparison of CO and H coverages on Ni and titania-containing Ni as a function of temperature at a total pressure of 760 Torr and an H_2/CO ratio of 4.

The H-adatom coverage on clean Ni shows an approximately Arrhenius dependence at low temperature, with a limiting apparent activation energy of ca. 90 kJ-mol^{-1}. This value gradually decreases to lower values at higher temperatures. Qualitatively, this calculated dependence parallels the experimentally-determined temperature dependence of the methanation reaction rate over polycrystalline (19,20) and single crystal nickel (16,17). It appears, therefore, that a controlling factor in the observed methanation kinetics is the surface coverage in the chemisorbed layer. This conclusion is compatible with mechanisms which involve H-adatom participation in the rate-determining step. This would include both surface carbon hydrogenation and hydrogen-assisted CO dissociation.

A more complete description of the reactant and intermediate surface concentrations was developed using a Langmuir-Hinshelwood model involving the following elementary steps over a nickel surface:

$$H_2(g) + 2* \underset{k_d^H}{\overset{k_a^H}{\rightleftharpoons}} 2H(a)$$

$$CO(g) + * \underset{k_d^{CO}}{\overset{k_a^{CO}}{\rightleftharpoons}} CO(a)$$

$$CO(a) + * \xrightarrow{k_1} C(a) + O(a)$$

$$C(a) + H(a) \xrightarrow{k_2} CH(a)$$

Remaining steps were assumed to be kinetically insignificant. For negligible surface oxygen concentration, the following rate expression can be derived:

$$r_{CH_4} = \frac{k_1 K_A^{CO} P_{CO}}{\left[1 + K_A^{CO} P_{CO} + \frac{k_1 K_A^{CO} P_{CO}}{k_2 (K_A^H P_{H_2})^{1/2}} + (K_A^H P_{H_2})^{1/2} \right]^2}$$

where K_A^{CO} and K_A^H are the equilibrium constants for CO and hydrogen adsorption, respectively. Note that a rate-determining step was not assumed in this mechanism; instead the methanation rate is determined by a balance between the surface carbon formation and removal rates as suggested by Goodman and coworkers (17).

Using activation energies for CO dissociation and carbon hydrogenation reported in the literature (i.e., 230 and 145 kJ/mol, respectively), calculations showed that changes in heats of adsorption of the reactants H_2 and CO on Ni significantly alter methanation behavior. Turnover frequencies for the Ni and TiO$_x$/Ni surfaces are plotted in an Arrhenius fashion in Figure 9 for a total pressure of 120 Torr and an H_2/CO ratio of 4. The plots show that the presence of titania adspecies increases the methanation rate one to two orders of magnitude for all temperatures below ca. 650 K.

This increased activity of the TiO_x/Ni surface is a direct consequence of corresponding higher H-adatom surface coverages on the titania-containing Ni surface. Figure 10 compares the calculated coverages of C, H, and CO for the clean and titania-containing Ni surfaces. The figure shows that the major difference between these two surfaces is that titania weakens CO to a sufficient extent that significantly greater amounts of hydrogen (ca. > two orders of magnitude at 500 K) are adsorbed. The figure also reveals that the gradual increase in apparent methanation activation energy as reaction temperature is decreased is due to a shift in the most abundant surface intermediate from C to CO. Studies using isotopic transient tracing have given results in qualitative agreement with the model, showing that at low temperatures the surface is mostly covered with CO, while at higher temperatures carbidic carbon and partially-hydrogenated carbonaceous intermediates (CH_x) dominate (21-23). We also note that qualitatively, the predicted dependence of apparent activation energy on reaction temperature has been experimentally observed (20,24).

It is unclear how the calculated shifts in C and H surface coverages could lead to the greater selectivities to higher hydrocarbons typically observed for titania-supported Ni catalysts (25). A possible explanation lies in the work of Kelley and Semancik (26), who showed that higher hydrocarbon formation on Ni was strongly dependent on CO partial pressure. The authors concluded that CO weakly adsorbed at hollow or bridging sites (which are favored over atop sites for titania-containing surfaces) participated in chain growth reaction steps.

Conclusions

The effects of titania adspecies on Ni are short-ranged and can largely be explained in geometric terms. It is suggested that titania adspecies block CO adsorption at strongly-bound linear sites and enhance adsorption at more-weakly-bound multifold sites. In contrast to CO adsorption, hydrogen adsorption is strengthened by the presence of titania. This may be due to direct bonding of hydrogen adatoms to titania. Carbon monoxide and hydrogen adsorption behavior cannot be consistently explained in terms of electronic interaction of titania with nickel. Electronic effects, although possibly present, appear to be of secondary importance to geometric and structural factors.

Calculations show that changes in CO and hydrogen adsorption strengths caused by titania adspecies on nickel lead to significantly greater H-adatom coverages on Ni sites under steady-state methanation conditions. This shift in the competitive nature of hydrogen and CO adsorption could be responsible for the higher methanation reaction rates typically observed for nickel supported on titania relative to nickel on silica or alumina. Importantly,

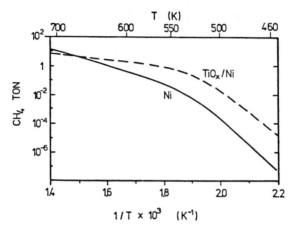

Figure 9. Comparison of the turnover frequency (TOF) for methane synthesis over a clean Ni and a titania-containing Ni surface based on model calculations at a total pressure of 120 Torr and an H_2/CO ratio of 4.

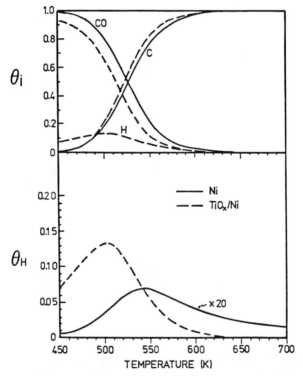

Figure 10. Surface coverages for clean Ni and titania-containing Ni as a function of temperature corresponding to the curves of Figure 9.

the activity enhancement predicted by this model could be realized
without invoking a change in mechanism or rate-determining step and
without assuming the existence of special sites for CO bond breaking
at the interface between Ni and titania.

Literature Cited

1. Raupp, G. B.; Dumesic, J. A. J. Phys. Chem. 1984, 88, 660.
2. Raupp, G. B.; Dumesic, J. A. J. Phys. Chem. in press.
3. Kiskinova, M.; Goodman, D. W. Surf. Sci. 1981, 108, 64.
4. Takatani, S.; Chung, Y.-W. J. Catal. 1984, 90, 75.
5. Madix, R. J.; Thornburg, M.; Lee, S. B. Surf. Sci. 1983, 133, L447.
6. Gland, J. L.; Madix, R. J.; McCabe, R. W., DeMaggio, C. Surf. Sci. 1984, 143, 46.
7. Anderson, S. Solid State Commun. 1977, 21, 75.
8. Tracy, J. C. J. Chem. Phys. 1972, 56, 2736.
9. Prichard, J. Surf. Sci. 1979, 79, 231.
10. Biberian, J. P.; Van Hove, M. A. Surf. Sci. 1982, 118, 443.
11. Campuzano, J. C.; Greenler; R. G. Surf. Sci. 1979, 83, 301.
12. Goodman, D. W.; Yates, J. T., Jr.; Madey, T. E. Surf. Sci. 1980, 93, L135.
13. Mitchell, G. E.; Gland, J. L.; White, J. M. Surf. Sci. 1983, 131, 167.
14. White, J. M. J. Phys. Chem. 1983, 87, 915.
15. Koel, B. E.; Peebles, D. E.; and White, J. M. Surf. Sci. 1983, 125, 709.
16. Kelley, R. D.; Goodman, D. W. Surf. Sci. 1982, 123, L743.
17. Goodman, D. W.; Kelley, R. D.; Madey, T. E.; Yates, J. T., Jr. J. Catal. 1980, 63, 226.
18. Vannice, M. A.; Twu, C. C.; Moon, S. H. J. Catal. 1983, 79, 70.
19. Polizotti, R. S.; Schwarz, J. A. J. Catal. 1982, 77, 1.
20. Kelley, R. D.; Madey, T. E.; Revesz, K.; Yates, J. T., Jr. Appl. of Surf. Sci. 1978, 1, 266.
21. Happel, J.; Suzuki, I.; Kokayeff, P.; Fthenakis, V. J. Catal. 1980, 65, 59.
22. Happel, J.; Cheh, H. Y.; Otarod, M.; Ozawa, S.; Severdia, A. J.; Yoshida, T.; Fthenakis, V. J. Catal. 1982, 75, 314.
23. Biloen, P.; Helle, J. N.; van den Berg, F. G. A.; Sachtler, W. M. H. J. Catal. 1983, 81, 450.
24. Weatherbee, G. D.; Bartholomew, C. H. J. Catal. 1982, 77, 460.
25. Vannice, M. A.; Garten, R. L. J. Catal. 1979, 56, 236.
26. Kelley, R. D.; Semancik, S. J. Catal. 1983, 84, 248.

RECEIVED September 17, 1985

5

Support Effects Studied on Model Supported Catalysts

R. A. Demmin, C. S. Ko, and R. J. Gorte

Department of Chemical Engineering, University of Pennsylvania, Philadelphia, PA 19104

Support-metal effects have been studied on model
catalysts in which titania was deposited on Pt,
Rh, and Pd foils and niobia was deposited onto a
Pt foil. These samples were then characterized
by Auger electron spectroscopy (AES), temperature
programmed desorption (TPD), and methanation
kinetics. Results indicate that a partial oxide
of titanium can form an even layer on each metal
and this layer causes the complete suppression of
CO and H_2 adsorption at a titania coverage corresponding to 1×10^{15} oxygen atoms/cm^2. Similar
results are observed for niobia on Pt. Methanation
rates on Pt show that the niobia and titania-
covered foils exhibit rates which are identical
to rates on actual Pt/TiO_2 catalysts, even though
no CO or H_2 adsorption is observed in TPD. These
results are used to explain the reasons behind
the strong metal-support interactions observed
for niobia and titania.

The use of titania and niobia as supports for a catalytic metal
can dramatically alter the properties of that metal.([1-7])
Recently, evidence has been presented by several researchers
that these effects in the case of titania are due to the
presence of a partially reduced oxide layer on top of the metal
particles.([8-20]) For example, we have shown that a Pt foil
with a titania overlayer exhibits identical adsorption and
reaction properties to an actual titania-supported Pt
catalyst.([8]) Others have come to the same conclusion using
other types of samples.

   In this paper, we will discuss results with titania on Rh
and Pd. While differences are observed in the results for

0097-6156/86/0298-0048$06.00/0
© 1986 American Chemical Society

these different metals, there are significant similarities
between them. Since the titania layer on each of these metals
appears to be bonded to the metal surfaces,(18) the differences
that are observed for each metal can be explained in terms of
specific interactions between the oxide layer and each metal.
In addition to these results, we will discuss the methanation
rates we obtained from a Pt foil with a niobia overlayer. The
niobia-covered foil exhibited identical rates to the titania-
covered foil, indicating that niobia support effects are caused
by overlayers similar to those found with titania.

Results

The experimental techniques and apparatus have been described
in detail elsewhere.(8,9) Titania and niobia were deposited
onto the clean metal foils by vaporizing either a Ti-Ta alloy
wire or a niobium wire in 10^{-7} Torr H_2 at 800K. Auger electron
spectroscopy (AES) was used to determine the oxide coverages,
and temperature programmed desorption (TPD) was used to determine
the effect of the oxide layers on the adsorption of CO and H_2.
Methanation rates were measured in a side chamber which allowed
the sample to be characterized by AES before and after rates
were measured. All rates were measured with 100 Torr CO and
400 Torr H_2, and conversion to methane was always kept less
than 1%.

Titania-Covered Pt, Rh, and Pd

AES results for the clean and the titania-covered Pt, Rh, and
Pd surfaces are discussed in detail elsewhere.(9) Relative
coverages for each of these surfaces were determined by comparing
the ratio of the Ti(385eV) peak height to either the Pt(238eV)
peak, the Rh(302eV) peak, or the Pd(330eV) peak. In all cases,
the O(508eV) peak was found to be very sensitive to electron
beam irradiation; therefore, this region of the spectrum was
always taken first before measuring the rest of the spectrum.
Because chemisorbed oxygen is not electron beam sensitive on
any of these surfaces, we believe this is evidence that the
oxygen on each of these surfaces is associated with the titanium.
    The ratio of the O(508eV) to the Ti(385eV) peak was constant
for all titania coverages on each metal, although this ratio
was different for each metal. These ratios were 0.6 for Pt, 1.1
for Rh, and 1.2 for Pd. The ratio for bulk TiO_2 is 2.5,(21)
indicating that the species on each metal is likely to be only
a partial oxide. X-ray photoelectron spectroscopy (XPS) of the
titania layer on Pt indicated that this oxide layer has a stoichiometry close to that of TiO.(18)
    The thermal stability of the titania layer was found to
vary with the metal substrate. On Pt, this oxide layer was
found to diffuse into the bulk at high temperatures and segregate
back to the surface at low temperatures.(19) On Rh, the titania
layer did not disappear from the surface at high temperatures;
however, we did observe evidence for the migration of oxide

into the bulk metal. We found that we could remove the oxide
layer on Rh by ion bombardment but, when this sample was annealed
to high temperatures in vacuum, titania would reappear from the
bulk. For both Pt and Rh, therefore, we have found evidence
that a titanium oxide species is mobile in the bulk metal.

The results for Pd were different. On Pd, we found that
the titanium oxide decomposed at 1100K. At this temperature the
titanium apparently diffuses into the bulk and forms an alloy.
Titania can be forced back to the surface only by annealing the
sample at high temperatures in oxygen.

The only effect of the titania layers on the adsorption of
CO and H_2 on these surfaces, as measured by TPD, was to decrease
the amount of each gas that could adsorb. In all cases, we
performed H_2 adsorption at 80K and CO adsorption at 295K. On
Pt, H_2 exhibited three desorption states at 220, 300, and 370K
and CO exhibited two states at 400 and 520K. For Rh, H_2 desorbed
in two states at 120 and 270K, and CO desorbed in one broad
state at 480K. For Pd, desorption occurred at 160 and 300K for
H_2 and at 470 for CO.

No new desorption features were ever observed due to the
presence of titania. In all cases except for H_2 desorption
from Pd, the shape of the TPD curve remained unchanged and the
relative coverages for each desorption feature remained the
same. Titania only affected the total amount of gas that could
adsorb, and there was no evidence for interactions between the
oxide and the chemisorbed gases. For H_2 adsorption on Pd, the
desorption feature at 160K was affected more dramatically in
that very small coverages of titania tended to suppress this
peak preferentially. There is evidence in the literature that
the 160K peak is related to a surface reconstruction(22);
therefore, this preferential suppression should not be interpreted
as evidence for interaction between the oxide and the H_2.

The amounts of CO and H_2 that could adsorb on each surface
decreased linearly with the titania coverage, showing that
titania forms an even layer on these metals. The coverages of
titania necessary to completely suppress the adsorption of H_2
and CO have been determined using calibrations determined for
oxygen on each of these metals. For each metal, this titania
coverage corresponded to approximately 10^{15} oxygen atoms/cm^2, a
coverage close to that expected for a close-packed monolayer.(9)
This indicates that chemisorption suppression is likely due to
the titania physically blocking the surface and not to any
electron transfer mechanism.

Niobia – Covered Pt

Niobia was deposited on the Pt surface in the same manner
described for titania. Evaporation was carried out in 10^{-7}
Torr O_2 to ensure complete oxidation and the sample was then
annealed to high temperatures to remove oxide multilayers. As
with titania, we found that the niobia could diffuse into bulk
Pt at high temperatures and tended to segregate back to the
surface at low temperatures. The only effect niobia had on the

adsorption of CO and H_2, as measured by TPD, was to decrease
the coverages. Complete suppression of CO and H_2 adsorption
occurred at a niobia coverage close to $10^{15}/cm^2$, as measured by
the O(508eV)/Pt(238eV) ratio.

Methanation Rates

Methanation rates for Pt with 400 Torr H_2 and 100 Torr CO are
shown in Figure 1. The rates for the niobia and titania-
covered surfaces were measured with an oxide coverage which
completely suppressed H_2 and CO adsorption. As discussed
earlier, our experimental apparatus allowed us to examine the
surface by AES both before and after reaction. On the clean
surface, we found that no impurities, including carbon, were
present following reaction. Also, below 750K, the titania and
niobia layers were unaffected by reaction conditions.
Therefore, the rates presented in this figure are not due to
any impurity effects or changes in oxide coverage during
reaction.

What these rates show is that, over the temperature range
of practical interest, the rates on the oxide-covered surfaces
are considerably higher than on the clean surface. Also, the
activation energy on the oxide-covered surfaces is considerably
lower than on the clean surface, going from 30.2 kcal/mole on
the clean surface to 18.9 kcal/mole on the oxide-covered
surfaces. We believe this lower activation energy is a result
of changes in the energetics of the reaction pathways and is
not responsible for the increased rates on the oxide-covered
surfaces, since no such activation energy changes are observed
for actual supported catalysts.

It is interesting to compare our reaction rates to those
reported for actual supported catalysts.([2],[3]) The rates for
our oxide-covered surfaces are almost identical to those
reported in the literature for titania-supported catalysts.
This is very strong evidence that our model catalysts are
similar to actual catalysts and are reproducing the important
features of titania-supported Pt. The rates for the clean Pt
surface are in the range reported for Pt/Al_2O_3 but are not
identical to that surface. These rates are considerably
different from those on Pt/SiO_2. We believe these variations
may be due to crystallographic effects. Differences between
Pt/Al_2O_3 and Pt/SiO_2 catalysts have been assigned to
morphological effects by other workers, and this may also be
responsible for variations in the rates on the Pt foil.([23])

It is also interesting to note that the rates on the
niobia-covered and the titania-covered surfaces are identical
within the accuracy of our experiment. Since titania and niobia
are chemically different, this similarity suggests that the
effect these oxides have on the activity is not due to electronic
interactions. While XPS results suggest some bonding occurs
between the adsorbed titania and the metal, one would expect
the amount of electron transfer to be different for niobia;
therefore, we do not believe that electron transfer is the

predominent factor for the enhanced activities that we observe.
Rather, we believe that the oxide layers may play a geometric
role in allowing CO to approach the metal only up to a certain
distance or at a certain angle. No firm reasons for the effect
of the oxide layers can be given since the mechanism for methanation
on Pt is not known in detail.

Summary

We have presented evidence that the strong metal-support
interactions observed with titania and niobia are due to an
oxide layer over the metal catalyst. This layer interacts
chemically with the metal, as evidenced by the fact that the
titania layers on Pt, Rh, and Pd do have slightly different
properties. The fact that the methanation rates for a titania-
covered and a niobia-covered Pt foil are identical indicates
that the reason for enhanced methanation activity on the oxide-
covered surface is likely due to geometric and not electronic
considerations.

Acknowledgments

This work was partially supported by the NSF through the MRL
Program, Grant #DMR-7923647. Some equipment was provided by
the Research Corporation, Grant #9583.

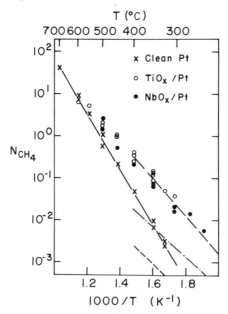

Fig. 1   Methanation rates on a Pt foil. The dashed lines
          indicate rates reported for — — — Pt/TiO$_2$, - · - ·
          Pt/Al$_2$O$_3$, - - - - Pt/SiO$_2$. The rates for the titania
          and niobia-covered foils were measured with an oxide
          coverage high enough to completely suppress CO and H$_2$
          adsorption.

Literature Cited

1. Tauster, S. J.; Fung, S. C. and Garten, R. L.
   JACS 100, 170 (1978).
2. Vannice, M. A.; Twu, C.C. and Moon, S.H., J.
   Catal. 79, 70 (1983).
3. Vannice, M. A. and Twu, C. C., J. Catal. 82 213 (1983).
4. Vannice, M. A., J. Catal. 74, 199 (1982).
5. Ryndin, Y. A.; Hicks, R. F.,; Bell, A. T. and Yermakov, Y.
   J. Catal. 70, 287 (1981).
6. Wang, S. Y.; Moon, S.H., and Vannice, M. A. J. Catal. 71,
   167 (1981).
7. Ko, E. I.,; Hupp, J. M., and Wagner, N. J., J. Catal. 86,
   315 (1984).
8. Demmin, R. A.,; Ko, C. S., and Gorte, R. J., J. Phys. Chem.
   89, 1151 (1985).
9. Ko, C.S. and Gorte, R.J., Surface Sci., in press.
10. Simoens, A. J.,; Baker, R. T. K.; Dwyer, D. J., Lund, C. R.
    F. and Madon, R. J., J. Catal. 86, 359 (1984).
11. Belton, D. N., Sun, Y. M. and White, J. M., J. Phys. Chem.
    88, 5172 (1984).
12. Sadeghi, H. R. and Henrich, V. E., J. Catal. 87, 279 (1984).
13. Raupp, G. B. and Dumesic, J. A., J. Phys. Chem. 88, 660
    (1984).
14. Santos, J.; Phillips, J., and Dumesic, J. A., J. Catal. 81,
    147 (1983).
15. Spencer, M. S., J. Phys. Chem. 88, 1046 (1984).
16. Resasco, D. E. and Haller, G. L., J. Catal. 82, 279 (1983).
17. Vannice, M. A. and Sudhaker, C., J. Phys. Chem. 88, 2429
    (1984).
18. Greenlief, C. M.; White, J. M.; Ko, C. S. and Gorte, R. J.,
    J. Phys. Chem., in press.
19. Ko, C. S., and Gorte, R. J., J. Catal. 90, 32 (1984).
20. Ko, C. S. and Gorte, R. J., Surface Sci. 151 (1985) 296.
21. Davis, G. D.; Natan, M., and Anderson, K. A., Appl. of
    Surface Sci. 15, 321 (1983).
22. Cattania, M. G.; Penka, V.; Behm, R. J.; Christmann, K. and
    Ertl, G., Surface Sci. 126, 382 (1983).
23. Wong, S. S.; Otero-Schipper, R.H., Wachter, W. A.; Inoue,
    Y; Kobayashi, M., Butt, J. B., Burwell, R. L. and Cohen, J.
    B., J. Catal. 64, 84 (1980).

RECEIVED October 17, 1985

6

Evidence for the Migration of MnO upon Reduction of Ni–MnO$_x$ and Its Effects on CO Chemisorption

Y. W. Chung and Y. B. Zhao

Department of Materials Science and Engineering, Northwestern University, Evanston, IL 60201

Model Ni/MnO$_x$ catalysts were prepared by 150-Å nickel deposition onto an oxidized manganese disk under ultra-high vacuum. The surface composition of the catalyst was followed as a function of time at 500°K in ultra-high vacuum. Rapid diffusion of reduced manganese oxide onto the nickel surface was observed. This oxide migration phenomenon was observed for a wide range of metal/TiO$_2$ catalysts reduced at $\gtrsim$ 700°K and is presently accepted as the key step in the induction of strong metal-support interaction (SMSI). The present observation appears to be the first case in which SMSI can be induced at normal (500°K) catalyst reduction temperatures. High resolution electron energy loss spectroscopy (EELS) experiments of CO on Ni/MnO$_x$ reduced at 500°K showed dramatic reduction of the C-O stretching frequency of adsorbed CO. These EELS results are discussed in light of similar data on Ni/TiO$_2$ and infra-red data on Rh/MnO$_x$.

After the 1978-publication of Tauster and coworkers on CO and H chemisorption suppression on titania-supported group VIII catalysts reduced at 500°C ($\underline{1}$), many research studies were performed to elucidate the mechanism of this so-called strong metal-support interaction (SMSI) phenomenon. The first direct clue came from the work of Dumesic and Haller. Work by Dumesic and coworkers ($\underline{2}$) on ammonia symthesis using Fe/TiO$_2$ showed that SMSI behavior persists even for iron particles with diameter exceeding 200-Å. This finding excludes the delocalized charge transfer model as a viable explanation of SMSI since such charge accumulation will be completely screened out in one atomic spacing by the metal. Resasco and Haller ($\underline{3}$) found that the ethane hydrogenolysis activity of Rh/titania decreases as the square-root of reduction time, the drop being faster with higher reduction temperature. Both groups proposed that during high temperature reduction to induce SMSI, titania diffuses to the metal surface.

We provided a direct confirmation of this oxide migration model
by starting with a single crystal Ni(111), depositing onto it controlled amounts of reduced titania, and performing CO chemisorption
and CO hydrogenation reactions on such a surface (4). Such a TiO$_x$/
Ni(x<2) surface exhibits all the standard characteristics of SMSI,
viz. suppressed CO chemisorption, higher methanation activity, and
shift of product distribution to heavier molecular hydrocarbons.
Additional Auger and profiling experiments on a 120-Å Ni/TiO$_2$ model
catalyst showed the rapid migration of reduced titania upon reduction at 700°K (5,6). Migration of titania was also found for Rh/
TiO$_2$ (7,8) and Pt/TiO$_2$(9). It is now well established that oxide
migration onto the metal surface is the key process leading to SMSI.
       On the other hand, catalytic effects of this surface titanium
suboxide are not so clear. There is evidence in the literature (9)
and in this Symposium (10) that geometric effects are dominant in
Pt. But in the case of Ni, the bonding of CO and H$_2$ is altered (6,
11). Sachtler (12) suggested that the oxophilic nature of reduced
titanium may weaken the C-O bond upon CO chemisorption near such
sites. If this is true, other oxophilic ions should show similar
characteristics. In this paper, we report Auger and EELS studies of
Ni/MnO$_x$ model catalysts prepared by vacuum evaporation of Ni onto an
oxidized manganese foil and subsequent reduction at 500°K.

Experimental Section

All studies were done in an ultrahigh vacuum chamber equipped with
an Auger spectrometer, a high resolution EELS spectrometer, sputtering, Ni evaporation, standard gas and specimen handling facili-
ties. The manganese specimen used in this study is a 1 cm^2- 0.5 mm
thick polycrystalline disk purged of bulk impurities by a series of
sputtering and annealing cycles. A MnO$_x$ (x in the range of 0.9 -
1.2 using PHI Auger Handbook data) film was formed by heating to
300°C under 3x10^{-6} Torr oxygen for 5 to 15 minutes, followed by a
150-Å Ni deposition. The specimen temperature was kept at ambient
during evaporation. The resulting specimen was annealed at 500°K
under ultrahigh vacuum (UHV). Auger profiles of Mn, Ti, and O were
obtained by standard sequential argon ion sputtering and Auger analysis. HREELS studies of the C-O stretching frequency of adsorbed
CO on such surfaces were done at an electron impact energy of 2 eV
(uncorrected for work function differences) and an overall resolution of 15 meV after a CO exposure of 2x10^{4}L.

Results and Discussions

Auger Results

Figure 1 shows the evolution of the Ni(848 eV), Mn(589 eV), and O
(503 eV) Auger peak intensities as a function of reduction time at
500°K in UHV on a 150-Å Ni/MnO$_x$ model catalyst. All Auger peak intensities were converted to atomic concentration using data from
the PHI Auger Handbook. The rapid increase of manganese and oxygen
Auger intensity is apparent. If all the oxygen is associated with
manganese, the stoichiometry corresponds to roughly MnO$_{0.4-0.7}$.

Fig. 2 shows the sputter profile of the model catalyst after
a 60 minute reduction at 500°K. Before reduction, one does not
observe the presence of oxygen and manganese on the nickel surface.
After reduction, the segregation of oxygen and manganese is apparent.

There are similarities and differences between the above findings and those of Ni/TiO$_2$ (6). In both cases, we observed the migration of the support oxide onto the Ni surface. However, the
rates are very different. In order to observe significant segregation of Ti in Ni/TiO$_2$ in a reasonable amount of time, reduction
temperatures $\gtrsim 700^{\circ}$K must be used. However, in the case of Ni/
MnO$_x$, massive segregation occurs within a few minutes at a reduction temperature of 500°K. A rough estimate of the diffusion coefficient D can be made as follows. A substantial amount of Mn
was observed on the Ni surface (thickness 150Å) after about 100
seconds. From the equation ℓ^2 = 2Dt, where ℓ is the diffusion
distance (150 Å) and t the diffusion time (100 sec.) D was calcu-
lated to be about 1×10^{-14}cm^2/sec. at 500°K. As a comparison, the
diffusion coefficient for the case of Ti was estimated by extrapolation of our previous data (6) to be about 4×10^{-19}cm^2/sec. at
500°K --- Mn diffuses more than 100 times farther than Ti under
identical reduction conditions.

Apart from kinetics, the fact that reduced manganese oxide
(MnO) appears on the surface of Ni implies that the bond energy
between MnO and Ni must exceed the lattice Madelung energy of the
oxide. Since the electronic configuration of Mn in this case is
3d^5, i.e. a half-filled d shell, this suggests that the d-d bonding between Ni and Mn may be a crucial driving force for the oxide
segregation. Further, such d-d bonding may also affect the chemistry occuring on Ni sites as well. There is now EXAFS evidence
in the case of Rh/TiO$_2$ of such direct Rh-Ti bonding (13)

HREELS Results

Fig. 3 shows EELS spectra near the C-O stretch region of Ni/MnO$_x$
catalyst reduced at 500°K for zero and 10 minutes followed by 2×10^4
L CO exposure. With no reduction, the major C-O stretch is located
at 1900 cm^{-1}, with a shoulder around 2040 cm^{-1}. These features are
quite representative of CO chemisorption on Ni (6) and are assigned
to CO chemisorbed on two-fold and on-top sites respectively. However, after 10 minutes of reduction, evolution of a new peak at
1750 cm^{-1}is apparent. From the correlation of the appearance of
these features and Mn Auger signals, it is clear that these new C-O
bands are due to CO chemisorption on or around the MnO sites.

Lowering of the C-O stretch frequency has also been observed on
Ni/TiO$_2$. More recently, Ichikawa and coworkers (14) found that CO
chemisorption on Rh catalysts promoted with Mn results in new CO
bands at 1780 and 1530 cm^{-1}, similar to what we report here.

Based on these observations, we believe that the close proximity between nickel and surface manganese oxide as a result of d-d
bonding makes it possible for the two ends of CO to be activated by
them simultaneously. The strong bonding and the resulting lowering
of surface free energy are the primary driving forces for the migration of manganese oxide. The end result is similar to that of a

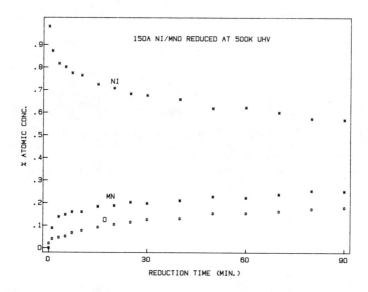

Figure 1.   Auger intensity of Mn, 0, and Ni at 589 eV, 503 eV, and 848 eV respectively, as a function of reduction time.

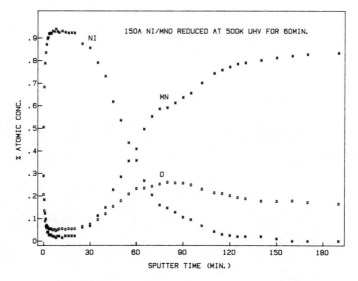

Figure 2.   Sputter profile of a 150Å Ni/MnO$_x$ model catalyst after 60 minutes of reduction at 500°K.

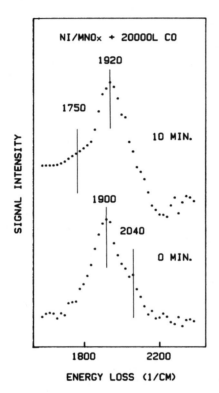

Figure 3. High resolution EELS spectra of Ni/MnO$_x$ after 500°K reduction for various times, followed by 2x10^4 L CO exposure. Lower curve - zero minute, upper curve - 10 minutes.

catalyst promoter/poison, except that such a promoter/poison is derived inherently from the oxide support.

Summary

Model Ni/MnO$_x$ catalysts have been prepared by deposition of 150Å of
Ni onto an oxidized Mn disk. Subsequent reduction at 500°K in ultrahigh vacuum results in rapid segregation of MnO onto the Ni surface. The segregation rate is much faster than that of Ti under
identical conditions. HREELS studies reveal new C-O stretching
bands at lower frequencies, comparable to what have been observed
on Rh promoted with Mn. We assign these bands to be due to CO chemisorbed on or near MnO sites.

Acknowledgments

Acknowledgment is made to the Donors of the Petroleum Research Fund,
administered by the American Chemical Society, for support of this
research. We thank Professor Sachtler and Dr. Ichikawa for many illuminating discussions.

Literature Cited

1.   Tauster, S. J., Fung, S. C., and Garten, R. L., *J. Amer. Chem.
     Soc.* 1978, 100, 170.
2.   Santos, J., Phillips, J., and Dumesic, J. A., *J. Catal.* 1983,
     81, 147.
3.   Resasco, D. E., and Haller, G. L., *J. Catal.* 1983, 82, 279.
4.   Chung, Y. W., Xiong, G., and Kao, C. C., *J. Catal.* 1984, 85,
     237.
5.   Takatani, S., and Chung, Y. W., *Applic. Surf. Sci.* 1984, 19,
     341.
6.   Takatani, S., and Chung, Y. W., *J. Catal.* 1984, 90, 75.
7.   Belton, D.N., Sun, Y. M., and White, J. M., *J. Phys. Chem.* 1984,
     88, 5172.
8.   Sadeghi, H. R., and Henrich, V. E., *J. Catal.* 1984, 87, 279.
9.   Ko, C. S., and Gorte, R. J., *J. Catal.* 1984, 90, 59.
10. Dwyer, D. J., Cameron, S.D., and Gland, J., *ACS Symposium on
     Metal-Support Interactions*, Miami, April 29 - May 1, 1985.
11. Raupp, G. B., and Dumesic, J. A., *J. Phys. Chem.* 1984, 88, 660.
12. Sachtler, W. M. H., *Proc. 8th Int. Cong. Catal.*, Berlin, 1984.
13. Haller, G. L., private communication (1985).
14. Ichikawa, M., private communication (1985).

RECEIVED September 30, 1985

7

Time Dependence of H_2 and O_2 Chemisorption on Rh–TiO$_2$ Catalysts

H. F. J. van 't Blik[1], P. H. A. Vriens, and R. Prins

Laboratory for Inorganic Chemistry, Eindhoven University of Technology, P.O. Box 513, 5600 MB Eindhoven, The Netherlands

A fast as well as a slow component was observed in the room temperature adsorption of H_2 and O_2 on Rh on rutile and anatase TiO$_2$ catalysts. Both components of the H_2 adsorption decreased in magnitude with increasing catalyst reduction temperature, while the fast O_2 adsorption increased in magnitude and the slow O_2 adsorption stayed constant. The fast H_2 adsorption proved to be due to chemisorption on the metal and the slow H_2 adsorption to spillover from the metal to the support. The fast O_2 adsorption was due to chemisorption on the metal as well as on the support, with the concurrent reoxidation of Ti^{3+} ions formed during H-spillover. The slow O_2 adsorption was caused by corrosive chemisorption of the metal particles.

Some years ago Tauster et al. reported that after a high temperature reduction of group VIII metals on supports like TiO$_2$, Nb$_2$O$_5$ and V$_2$O$_3$ the adsorption of H_2 and CO was suppressed, while O_2 adsorption was unaffected (1–3). They believed that these effects were caused by a special type of metal–support interaction and therefore called this interaction Strong Metal Support Interaction (SMSI). Catalytic studies showed that with the catalyst in the SMSI state many reactions were inhibited too. Thus hydrogenolysis of ethane (4, 5) and butane (5, 6), hydrogenation of ethene, benzene and styrene and dehydrogenation of cyclohexane (5, 6) and reforming of hexane (7) were much slower with catalysts reduced at 500°C than at 200 or 250°C. On the other hand CO hydrogenation was not inhibited at all, on the contrary for some metals on TiO$_2$ it was even somewhat better after high temperature reduction (on the basis of total metal present) (8–10).

[1] Current address: Philips Research Laboratories, P.O. Box 80000, 5600 JA Eindhoven, The Netherlands

The insensitivity of CO hydrogenation to reduction temperature may be connected to the fact that oxidants like O_2 and H_2O are capable of bringing the catalyst back from the SMSI state to the normal adsorption state. Thus Tauster et al. published that the H_2 adsorption capacity of a Pd/TiO$_2$ catalyst, which had been brought into the SMSI state by reduction at 500°C, was completely restored after oxidation at 400°C for 1 h and rereduction at 175°C ([1]). Baker et al. published that H_2O at 250°C for 1 h could restore the H_2 and CO adsorption capacities of a Pt/TiO$_2$ catalyst, although to a less extent than oxidation at 600°C ([11]). On the other hand, Mériaudeau et al. reported that H_2 adsorption as well as catalytic activities for hydrogenolysis and hydrogenation of TiO$_2$–supported Pt, Ir and Rh catalysts recovered after O_2 admission at room temperature and subsequent reduction at low temperature ([6]).

Explanations for SMSI have ranged from electronic theories, such as alloy formation ([1], [12]), metal–semiconductor interaction ([3], [6]) and metal–support cation charge transfer ([3], [13]), to geometrical theories like sintering, poisoning and covering ([5], [7], [14–19]). Today the electronic theory based on metal–support cation interaction and the geometrical covering theory are favoured most. In all theories the reducibility of the support plays an important role. In the electronic theories it is responsible for the form-ation of reduced cations on the surface of the support, e.g. Ti^{3+} on TiO$_2$. Xα scattered wave molecular orbital calculations on the interaction between a $(TiO_5)^{7-}$ cluster and a Pt atom indicated a strong electron transfer to the Pt atom, leading to $Pt^{-0.6}$. As a consequence there was a strong ionic bonding between Pt and the Ti^{3+} cluster ([13]). It is a pity that this calculation was done on such an unbalanced system, because the large difference in charge between $(TiO_5)^{7-}$ and Pt will always favour an electron flow towards Pt. A new calculation with compensating cationic charges around the $(TiO_5)^{7-}$ cluster would be of interest. Reducibility of the support and the formation of surface defects by dehydration at high temperatures are important ingredients in the model of Baker et al. ([11], [20]) and Huizinga and Prins ([21]). They suggested that the interaction between metal particles and domains of Ti_4O_7 under and around the metal particles might be strong, because of the metallic properties of the Magnelli phase type Ti_4O_7 suboxide. In the geometrical, covering theory the reducibility of the support, as well as the formation of surface defects by dehydration at high temperatures, are essential to explain the migration of reduced support species onto the metal particles ([5], [7], [14]).

The reducibility of TiO$_2$ has in recent years been studied with ESR and NMR techniques ([21–23]). In the course of our studies ([21]) we noticed that support reduction is a relatively slow process and that, as a consequence, hydrogen chemisorption on a metal on TiO$_2$ catalyst has a fast component, due to adsorption on the metal surface, and a slow component, due to spillover of H atoms from metal to support and subsequent support reduction. We have studied the time dependence of H_2 chemisorption, as well as that of O_2 chemisorption, in more detail and the results of this study are presented in this paper.

Experimental

Two Rh/TiO$_2$ catalysts were prepared by pore volume impregnation
of the support with an aqueous solution of RhCl$_3$.x H$_2$O (39 wt%,
Drijfhout). One catalyst was made with 0.99 wt% Rh on the rutile
modification of TiO$_2$ (Tioxide CLDD 1627/1, pore volume 0.57 ml g^{-1}
and surface area 20 m^2 g^{-1}), this catalyst will be further denoted
as Rh/R-TiO$_2$. The other catalyst with 1.00 wt% Rh was made with
anatase as the support (Tioxide CLDD 1367, pore volume 0.64 ml g^{-1}
and surface area 20 m^2 g^{-1}) and will be denoted as Rh/A-TiO$_2$.
It was checked that the main X-ray diffraction lines of the support,
as well as those of the catalysts prepared therefrom, were those of
rutile or anatase. Before use, both supports were washed twice with
distilled water, dried at room temperature and calcined at 500°C for
1 h to stabilise the surface area. After impregnation the catalysts
were dried at room temperature for 24 h and subsequently for 10 h at
120°C.
      Hydrogen and oxygen chemisorption measurements were performed
in a conventional glass system at 22°C. Before a H$_2$ chemisorption
measurement, the catalyst was reduced, or oxidized and reduced, at
temperatures and during times to be specified under Results. These
temperatures were reached with a heating rate of 5°C min^{-1}.
Reduction and oxidation took place in flowing hydrogen and oxygen,
respectively. After evacuation (10^{-2} Pa) at the same temperature
the cell was cooled down to room temperature and the chemisorption
of H$_2$ was started. The H$_2$ used was purified by passing through
a Pd diffusion cell. To check for activated chemisorption, in some
cases the chemisorption of H$_2$ was already started at the same
temperature as that of evacuation and the catalyst was then cooled
down to room temperature under H$_2$. Similar procedures were used
for oxygen chemisorption.
      All chemisorption experiments were single point measurements
at 8.10^4 Pa. By measuring the adsorption isotherms from 10^3-10^5 Pa
for a few catalysts, it was checked that a relative comparison of
the thus obtained chemisorption values was as justifiable as any
other method based on other measuring points, or on extrapolation
of measuring points to zero pressure, as advocated by Benson and
Boudart (24). No corrections were made for chemisorption on the
bare supports as such, because this was found negligible.

Results

H$_2$ chemisorption. Hydrogen chemisorption of a Rh/R-TiO$_2$ and a
Rh/A-TiO$_2$ catalyst were measured at 8.10^4 Pa as a function of time.
In Figure 1 the results for the Rh/R-TiO$_2$ catalyst are presented by
plotting the H/Rh ratios, calculated from the H$_2$ consumption and the
total amount of Rh present, as a function of ln t. This was done
because in that way nearly linear curves were obtained. Starting
from the impregnated and dried catalyst the first measurement was
carried out after reduction for 1 h and evacuation for 1 h at 215°C
(Figure 1A). Further experiments were carried out after a subsequent
prereduction at 520°C for 2 h, followed by a reoxidation at 140°C
for 1 h to break the SMSI state which results from the high temperature reduction. In the following section it will be shown that

140° C is sufficient to re-establish normal chemisorption behaviour.
After this treatment reductions and evacuations (for 1 h each) were
carried out subsequently at 200, 300, 400 and 500 °C, followed by
H_2 chemisorption measurements. The results of these measurements
are presented in Figures 1B, 1C, 1D and 1E, respectively. After the
last measurement (Figure 1E), the catalyst was further reduced at
500°C for 6 h, again an oxidation was applied at 125°C for 0.5 h,
the catalyst was reduced for 1 h and subsequently evacuated for 1 h
at 220°C, and then its H_2 chemisorption capacity was measured
again (Figure 1F). Further prolonged reduction at 220°C for 9 h,
followed by evacuation at 205°C for 2 h did not change the
chemisorption behaviour. The resulting H/Rh vs. ln t curve was
equal to curve F.

All measurements show a linear relationship between H_2
chemisorption (expressed as H/Rh ratio) and ln t (especially in the
first hours), a decrease in chemisorption with an increase in
reduction and evacuation temperature, and a decrease in the slope
of the H/Rh vs. ln t curves with increasing treatment temperature.
The results furthermore demonstrate that no equilibrium chemisorption was established within 18 h (ln t = 7) on this catalyst.

The decrease in H/Rh values between curves A and B demonstrates
that the treatment at 520°C had caused sintering. Further treatment
at intermediate temperatures followed by another treatment at 500°C
did not induce further sintering, or only a little bit (cf. curves
F and B). Therefore sintering of the Rh particles cannot be the
reason for the decreased H_2 chemisorption and decreased time
dependence (cf. curves B to E). Note that within the uncertainty of
the measurements there is no difference in slope between curves
A, B, and F, indicating that the slope is independent of the
dispersion of the metal.

To see if the time dependence of the chemisorption was caused
by a slow establishment of H_2 chemisorption at the Rh metal surface,
the measurements were also carried out for a 3.7 wt% Rh/SiO_2
catalyst (silica Grace S.P. 2–324.382, pore volume 1.2 ml g^{-1} and
surface area 290 m^2 g^{-1}) after reduction at 500°C. The H_2 chemisorption of this catalyst measured after a day under H_2 differed
only 10% from that measured directly after reduction, evacuation and
cooling (H/Rh= 0.46 and 0.42, respectively). Thus slowness of H_2
chemisorption onto the metal cannot be the reason for the time
dependence of the H_2 chemisorption on the $Rh/R-TiO_2$ catalyst.

That the time dependence of the H_2 chemisorption was anyway
caused by a slow attainment of equilibrium, was established by
performing the H_2 chemisorption measurement in a modified way.
Hydrogen was admitted to the evacuated catalyst at ca. 200°C and
subsequently the reactor was slowly cooled down to room temperature
and H_2 chemisorption was measured. In this case equilibrium was
quickly established and the chemisorption value was substantially
higher (about 20%) than that reached during room temperature
measurements for 18 h. Furthermore, since there was no difference
between the H_2 chemisorption measured after admission of H_2 at
190°C during 10 min and that measured after admission at 205°C
during 30 min, a temperature of about 200°C during 10 min seems
sufficient to quickly reach equilibrium.

Similar results as described above for the $Rh/R-TiO_2$ (rutile)

catalyst were obtained for the $Rh/A-TiO_2$ (anatase) catalyst, be it
that the slopes of the H/Rh versus ln t curves were smaller for Rh
on anatase than for Rh on rutile.

Breaking of SMSI by oxidation. In the literature it has been
stated that a metal–on–TiO_2 catalyst which is in the SMSI state,
can be brought back to the normal state by a treatment in oxygen.
Several conflicting statements have been published, however, about
the severeness of oxidation. Thus it has been said that keeping an
SMSI catalyst for some time in air would be sufficient (6), but in
another publication it has been claimed that high temperature
oxidation is absolutely necessary (1, 11). Furthermore, also water
vapour has been claimed to be able to break the SMSI state (11).
Taken all published information together, it seems clear that an
oxidative treatment is necessary, either with oxygen, water or any
other oxygen containing oxidant. The temperature and time needed
to completely break the SMSI state and to fully restore the normal
metal–on–support state are still unsettled, however. To study this
further we subjected a 0.99 wt% $Rh/R-TiO_2$ catalyst to a series of
successive reduction–oxidation–rereduction (200°C) treatments, with
evacuation after each reduction step, and measured the H_2 chemisorption after each treatment. The catalyst had been pretreated at
500°C for several hours to stabilise its rhodium dispersion. This
catalyst was precalcined at 150°C, (0.5 h), rereduced (1 h) and
evacuated (1 h) at 200°C; the subsequently measured H_2 chemisorption is presented as curve A in Figure 2. Subsequently the
catalyst was reduced again at 500°C for 1 h, oxidized in oxygen at
room temperature for 15 min (passivated), rereduced and evacuated
for 1 h at 200°C, and the H_2 chemisorption was measured. The
resulting curve B in Figure 2 demonstrates that the H_2 chemisorption had decreased, indicating that passivation for 15 min
cannot completely break the SMSI state. An extended passivation for
17 h left the chemisorption unaltered (curve C), proving that even
prolonged passivation could not fully restore the original chemisorption capacity. Oxidation at 150°C for 0.5 h (curve D) and
oxidation at 260°C for 0.5 h (curve E), however, could restore the
original chemisorption capacity and this shows that the restoration
of the catalyst from SMSI to normal state is an activated process.
Similar results were obtained for the $Rh/A-TiO_2$ catalyst.

O_2 chemisorption. Whereas the chemisorption of hydrogen is
suppressed in the SMSI state, chemisorption of oxygen still takes
place after a metal–on–TiO_2 catalyst has been reduced at high
temperature. No information is available, however, whether there is
a difference in O_2 chemisorption in SMSI and normal state and
whether the reduction temperature has any influence on this.
Therefore we looked into the O_2 chemisorption capacity of $Rh/R-TiO_2$
and $Rh/A-TiO_2$ catalysts at room temperature and 8.10^4 Pa after
successive reductions and evacuations (1 h each) at temperatures of
200, 245, 280, 350 and 500°C. The results are presented in Figures
3A, B, C, D and E, respectively. To ensure that the starting conditions were the same in all cases, before each reduction–evacuation
treatment and subsequent O_2 chemisorption experiment the catalyst
was oxidized at 150°C for 0.5 h because, as shown in the foregoing,

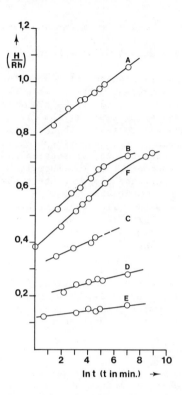

Figure 1. Influence of reduction temperature on the time dependence of the H_2 adsorption of 0.99 wt% Rh/TiO$_2$.

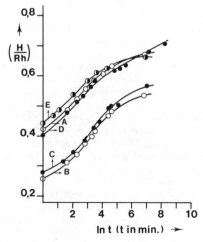

Figure 2. Influence of reoxidation temperature and time on the breaking of the SMSI state of 0.99 wt% Rh/R–TiO$_2$.

such an oxidation is sufficient to bring the catalyst back to the
normal state. Furthermore, to eliminate sintering effects the
catalyst was prereduced at 500°C for 2 h and reoxidized at 140°C
for 1 h.

We have presented the chemisorption results relative to the
total amount of rhodium present, that is as O/Rh values. The
results demonstrate that, just as for H_2 chemisorption, the O_2
chemisorption is dependent on time. In the first few hours there is
a more or less linear relationship with ln t, thereafter the
chemisorption levels off. The O_2 chemisorption is strongly
dependent on the pretreatment too, increasing reduction–evacuation
temperature leads to increasing O_2 chemisorption (Figure 3 curves
A to D). On the other hand the slope of the curves is not much
dependent on reduction temperature.

The results for the Rh/A-TiO$_2$ catalyst are qualitatively
similar to those of the Rh/R-TiO$_2$ catalyst, but quantitatively
the O_2 chemisorption is always somewhat higher for the Rh/A-TiO$_2$
catalyst and also the slope of the O/Rh vs. ln t curves is larger.

Measurement of the O_2 chemisorption of a reduced Rh/SiO$_2$
catalyst showed that this catalyst behaved similarly as the Rh/TiO$_2$
catalysts, with a very fast O_2 consumption, followed by a slow
O_2 uptake which had a linear behaviour with ln t. In contrast to
the Rh/TiO$_2$ catalyst, however, the fast O_2 consumption of the
Rh/SiO$_2$ catalyst proved to be independent of reduction temperature.

Discussion

H_2 chemisorption. Both Rh/R-TiO$_2$ and Rh/A-TiO$_2$ show a decrease
in H_2 chemisorption when the reduction and evacuation temperature
is increased, while at the same time the slope of the chemisorption
vs. ln t curve decreases. The decrease in H_2 chemisorption is of
course due to the gradual transition of the Rh particles into the
SMSI state. Whatever the explanation for this state, an electronic
interaction between metal particles and support or a covering of
the metal particles by the support, in this SMSI state the metal
particles are unable to adsorb H_2. The decreased slope of the
H/Rh–ln t curve can be explained in several ways, such as slow H_2
chemisorption on Rh because of an activated process, dependence on
metal dispersion, or an effect related to the support. The
experiments in which H_2 chemisorption was started around 200°C
proved that the time dependence is indeed due to a slow adsorption
at room temperature, but the experiment with Rh/SiO$_2$ showed that
there is no kinetic limitation in the H_2 chemisorption on the metal
part of the catalyst. In accordance with this conclusion, no effect
of rhodium dispersion on the time dependence of the H_2
chemisorption was observed for catalysts in the normal state
(cf. Figure 1 curves A, B and F).

The conclusion must be that the support is the cause of the
slow uptake of H_2 by the catalyst and this suggests that, in
addition to H_2 being bonded to the metal, H_2 must be bonded to
the support. It is of course well known that supports like TiO$_2$
and WO$_3$ can be reduced by hydrogen atoms which spill over from
metal particles to the support and it is not unlogic to presume
that spillover is an activated process, and thus is slow at room

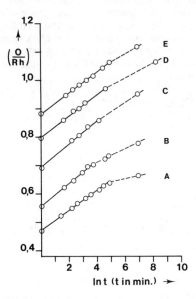

Figure 3. Influence of reduction temperature on the time dependence of the O_2 adsorption of 0.99 wt% Rh/R–TiO$_2$.

temperature (25, 26). At first sight, however, one encounters a
problem trying to explain our measurements with a spillover model.
Since all our catalysts had been reduced at temperatures of 200°C
and higher before the H_2 chemisorption at room temperature, spillover and support reduction should have been complete before the
start of the H_2 chemisorption measurements and no support
reduction would be expected during subsequent chemisorption measurements at room temperature. A closer look at the mechanism of support
reduction by spilled over hydrogen and at the conditions under which
the experiments were carried out, clarifies this contradiction. The
answer lies in the fate of the spilled over H atoms. Depending on
the temperature of reduction two reduction processes may take place,
one at low temperature

$$2 \text{ Ti}^{4+} + 2 \text{ O}^{2-} + H_2 \rightarrow 2 \text{ Ti}^{3+} + 2 \text{ OH}^- \tag{1}$$

and one at high temperature

$$2 \text{ Ti}^{4+} + \text{O}^{2-} + H_2 \rightarrow 2 \text{ Ti}^{3+} + H_2\text{O} \uparrow \tag{2}$$

Actually both processes can be composed of the following steps

$$H_2 \xrightarrow{\text{Rh}} 2 \text{ H} \tag{3}$$

$$2 \text{ H} + 2 \text{ Ti}^{4+} \rightarrow 2 \text{ H}^+ + 2 \text{ Ti}^{3+} \tag{4}$$

$$2 \text{ H}^+ + 2 \text{ O}^{2-} \rightarrow 2 \text{ OH}^- \tag{5}$$

$$2 \text{ OH}^- \rightarrow \text{O}^{2-} + H_2\text{O} \uparrow \tag{6}$$

The hydrogen atoms which are spilled over from the metal to the
support reduce the Ti^{4+} ions and the resulting protons are trapped
as OH^- ions (Equations 3, 4 and 5). At elevated temperatures the
protons are removed from the support surface by dehydration
(Equation 6). Huizinga and Prins demonstrated by means of ESR
measurements that Equations 3, 4 and 5 are reversible (21). Thus
reduction of Pt/TiO_2 and Rh/TiO_2 catalysts with H_2 at 300°C induced
a strong Ti^{3+} ESR signal, which disappeared after evacuation at
300°C, but stayed constant after evacuation at room temperature.
The authors concluded that desorption of H_2 by reversed spillover
and desorption from the metal was an activated process. These
phenomena were found reversible in the sense that when a catalyst,
which had been evacuated at 300°C, was exposed once more to H_2,
the strong Ti^{3+} ESR signal reappeared. This reversibility was not
observed when the reduction had taken place at 500°C, because in
that case the protons had left the TiO_2 surface as H_2O molecules
and Equations 5, 4 and 3 could no longer be reversed. At the same
time the titanium ions formed during high temperature reduction
were trapped in the 3+ oxidation state.
    In view of these ESR results, the explanation of the time
dependence of the H_2 chemisorption must be as follows. After
reduction in H_2 at 200°C indeed quite a bit of TiO_2 support
will be reduced, but during subsequent evacuation at 200°C all
Ti^{3+} ions will be reoxidized. As a consequence in the subsequent

H_2 chemisorption not only fast adsorption of H_2 on Rh may occur, but also a slow spillover and reduction of Ti^{4+}. On the other hand, after reduction at high temperature Ti^{3+} will be formed and the surface of the support will be dehydrated. Therefore during the following evacuation no reoxidation of the Ti^{3+} ions can take place. At the same time the rhodium has been changed into the SMSI state and therefore no H_2 chemisorption will take place at all. At intermediate reduction and evacuation temperatures there will be an intermediate behaviour and as a result of this the H_2 chemisorption values will decrease with increasing treatment temperature and so will their time dependencies.

The maximum amount of H_2 which can be chemisorbed onto the support of the Rh/TiO$_2$ catalyst (determined by admission of H_2 to the catalyst at 200°C) is of the same order of magnitude as the number of Ti^{3+} ions formed on Pt/TiO$_2$ ([21]) and Rh/TiO$_2$ ([27]) and detected by ESR. The agreement between H_2 chemisorption and ESR data means that all Ti^{3+} ions formed during reduction were indeed detected by the ESR technique and proves that these cations are not situated close together as nearest neighbours, because in that case antiferromagnetic coupling between adjacent Ti^{3+} ions would have led to a much decreased ESR signal.

The fact that the reduction of the support by hydrogen spillover is an activated process may be explained by hindered spillover from metal to support, by hindered reduction of Ti^{4+} ions by H atoms or by diffusion of H atoms over the support surface. Our results demonstrate that H_2 chemisorption equilibrium is quickly reached at 200°C, while the ESR results have shown that the Ti^{3+} ESR signal formed after reduction at 300°C could be decreased by two orders of magnitude by evacuation at 300°C, but not by evacuation at room temperature. Both the reduction and the reoxidation of the support therefore are activated processes, with similar activation energies. In the reduction step as well as in the reoxidation step, the oxidation state of the titanium cation changes. As a consequence also the cation radius and the Ti–O interatomic distances of the reduced or oxidized site will change and this will give rise to an activation energy barrier. After the initial reduction of Ti^{4+} cations in the immediate neighbourhood of a metal particle, further reduction of Ti^{4+} ions can only occur if the resulting Ti^{3+} ions diffuse away from the metal particle. The diffusion of an H atom over the very well dried, but not dehydroxylated, support surface can be described as hopping from one site to another, or alternatively as a synchronous hopping of a proton and an electron ([25], [28]). Each hopping consists of a simultaneous reduction of the receiving titanium cation and an oxidation of the leaving site, which may explain the similarity in activation energy for H_2 chemisorption on and desorption from the support.

Instead of the observed ln t dependence, a $t^{\frac{1}{2}}$ dependence of the H_2 chemisorption might have been expected. Such a relationship has indeed been observed up to t=5 min ([29]), but in our experiments it proved difficult to obtain accurate data at short measuring times. At the moment we do not have a physical explanation for the ln t dependence, but we note that a $t^{\frac{1}{2}}$ dependence is only expected under special conditions and that at large t, when the observed limited value of the slow adsorption is approached, the adsorption is neither described by a $t^{\frac{1}{2}}$ dependence, nor by a ln t dependence.

The proposed mechanism also explains why the slopes of the
H_2 chemisorption versus ln t curve for Rh on anatase were smaller
than those for Rh on rutile. For the Ti^{4+} cations at the rutile
surface are predicted to be more easily reduced than those at the
anatase surface (30).

The question may be posed if the slow reduction of the support
by spillover of H atoms from the metal particles takes place through
the bulk of the TiO_2 or on its surface only. The H_2 consumption
after exposure of H_2 to the Rh/R–TiO_2 catalyst at 200°C was found
to be independent of exposure time. Apparently only a limited amount
(0.3%) of the total available Ti^{4+} ions can be reduced at 200°C and
this suggests that this amount (which would constitute about 10% of
the surface Ti^{4+} ions) might well be located at the surface. In this
respect it is of interest to remark that Iyengar et al. observed
that the ESR signal of Ti^{3+} ions, formed after a 200°C reduction of
TiO_2, was quenched after O_2 admission at room temperature, while the
ESR signal obtained after a 500°C reduction remained (31).
Apparently Ti^{3+} ions formed at the surface by reduction only diffuse to the bulk at temperatures in excess of 200°C. In agreement
with this our O_2 chemisorption results also show that the major
part of the reoxidation of support Ti^{3+} ions is fast (vide infra).

In a recent publication Dumesic c.s. described adsorption and
desorption measurements of H_2 on Ni/TiO_2 and Pt/TiO_2 catalysts,
which showed that a larger amount of H_2 could be desorbed (after
15–20 h equilibration of these catalysts under about 40 kPa H_2)
than could be directly adsorbed (32). In agreement with our conclusions their explanation was that, apart from a fast H_2 adsorption
on the metal, hydrogen apparently also adsorbed slowly on the TiO_2
support via a spillover process from metal to support. These authors
noticed that the amounts of H_2 desorbed from the M/TiO_2 catalysts
in the SMSI state were in fair agreement with metal particle sizes
determined by X–ray line broadening and electron microscopy and
suggested that H_2 desorption could be used to estimate metal
particle sizes in M/TiO_2 catalysts in the SMSI state. In a former
publication (21) we have presented ESR evidence which shows that
indeed for Pt/TiO_2 the number of H atoms spilled over to the support
is about as large as the number of H atoms which can be adsorbed on
the Pt particles, provided they are not in the SMSI state.
Apparently hydrogen spillover can occur both in the SMSI and in the
normal metal–on–support state and is related to metal particle size.
Similar ESR results were obtained for Rh/TiO_2 catalysts (27). Our
present H_2 adsorption results for Rh on rutile are in conformance
with the ESR results, but the results for Rh on anatase indicate a
ratio of hydrogen adsorbed on the support to hydrogen adsorbed on
the rhodium below one. Judged from this limited experience it seems
that the amount of spillover hydrogen might give a semiquantitative
idea of the metal particle size in metal catalysts in the SMSI
state. Be that as it may, we would at the moment be reluctant to
generalize the above findings without further study and would much
rather support the method of breaking the SMSI state by reoxidation
and subsequently measuring the metal dispersion by H_2 adsorption.

O_2 chemisorption. Like the H_2 chemisorption, also the O_2 chemisorption had a slow component with a ln t behaviour in addition to a
fast uptake (cf. Figure 3). The fast component increased with

temperature of reduction, while the slope of the O/Rh vs. ln t curve
did not depend on reduction temperature. The Rh/SiO$_2$ catalyst had
a ln t type O$_2$ chemisorption too and thus the slow chemisorption
cannot be related to the TiO$_2$ support, but must be due to the
oxidation process of the rhodium particles. Temperature programmed
oxidation (TPO) measurements on rhodium catalysts with dispersions
around 0.5 have demonstrated that chemisorption and further
oxidation of the metal particles could be distinguished (33, 34).
TPO profiles of both Rh/TiO$_2$ catalysts indeed showed a low temperature peak due to chemisorption of O$_2$ and a second peak around 250°C.
Vis et al. attributed the occurrence of a second peak to a diffusion
limitation of Rh^{3+}and O^{2-} ions through the rhodium oxide layer
formed during the initial (corrosive) chemisorption (33, 34). For
rhodium catalysts with a very low dispersion they even detected a
third TPO peak at still higher temperature and attributed it to the
oxidation of the kernel of the very large metal particles. But, for
that assignment to be correct there should be a difference in the
diffusion mechanisms responsible for the second and third TPO
peaks, because otherwise only one peak would be present. In view of
the metal dispersions of about 0.4 present in the Rh/TiO$_2$ catalysts
after presintering, the second TPO peak around 250°C must be correlated with the slow O$_2$ chemisorption. The different diffusion
mechanisms then are a ln t type mechanism for the second TPO peak
and possibly a t$^{\frac{1}{2}}$ type mechanism for the third TPO peak. This seems
a very reasonable proposal since in metallurgy a ln t oxidation
process has indeed been observed for metal layer thicknesses of a
few tens of an Å, while a t$^{\frac{1}{2}}$ type oxidation is the normal
oxidation diffusion mechanism for thick metal layers (35, 36).

      In accordance with the explanation given for the slow O$_2$
chemisorption process, the slope of the O/Rh–ln t curves is
independent on reduction temperature. The catalysts had been
pretreated at 500°C and any reduction at a lower temperature thus
will not change the rhodium particle size, therefore the slopes
stay constant. The fact that the slope for the anatase based
catalyst was about 50% larger than that for the rutile catalyst is
also in agreement with this explanation. Judged from the initial
O/Rh values for the anatase and rutile catalyst, the Rh particles
in the Rh/A–TiO$_2$ catalyst had a dispersion which is 53% higher
than that of the Rh particles in the Rh/R–TiO$_2$ catalyst. Since
the slope of the O$_2$ chemisorption curves is caused by further
oxidation of the metal particles in a diffusion limited process,
the rate of this process will be dependent on the metal–oxygen
interfacial area, thus on the dispersion of the metal.

      Adsorption of O$_2$ on the Rh particles will contribute to the
fast initial O$_2$ chemisorption, but there must also be a contribution from the TiO$_2$ support to explain the increase in O$_2$ chemisorption with increasing reduction temperature of the Rh/TiO$_2$
catalysts. When increasing the reduction temperature Ti^{3+} cations
will be formed. Some of these cations will be reoxidized during the
subsequent evacuation at the same temperature (the reverse of
Equations 5, 4 and 3), but at increasing temperature more and
more of these cations will stay in the reduced form because of
Equation 6. As a consequence increasing numbers of Ti^{3+} ions
can be oxidized during O$_2$ chemisorption. The major part of this

reoxidation of Ti^{3+} ions by O_2 must be relatively fast, because
otherwise the support reoxidation would have shown up in the slow
O_2 chemisorption process in the form of increasing slopes at
increasing reduction temperatures. This fast reoxidation is not in
contradiction with the observed slow reduction of the corresponding
Ti^{4+} ions, considering that O_2 may directly chemisorb on Ti^{3+}
because both O_2 and Ti^{3+} have radical character, while H_2 can
only dissociatively adsorb on the metal and must be transported to
the Ti^{4+} ions via an activated spillover process.

SMSI. Finally we want to draw some conclusions which have a direct
bearing on the SMSI state. The results presented in Figure 2 illustrated that a catalyst which had been brought into the SMSI state
by reduction at 500°C and had subsequently been passivated by O_2
at room temperature and rereduced at 200°C, had the same slope of
the H_2 chemisorption vs. ln t curve as a catalyst which had only
been reduced at 200°C. In view of our discussion about the origin
of the slope, this suggests that the surface of the support is back
to normal after admission of O_2 at room temperature. This is in
agreement with our discussion of the O_2 chemisorption results,
which disclosed that reoxidation of Ti^{3+} ions is fast at room
temperature. In view of these conclusions it is difficult to
believe that suboxides like Ti_4O_7 are formed in the neighbourhood
of the metal particles to a great extent ([20], [21]), because such
suboxides are known to be stable at room temperature in air.
    The results presented in Figure 2 demonstrate that after
reduction at high temperature, passivation and rereduction at 200°C
the initial fast H_2 chemisorption only amounts to about 60% of the
value attained after low–temperature reduction only. Thus, although
passivation seems to have made at least part of the metal available
again for H_2 chemisorption, it is not sufficient to completely
restore the H_2 chemisorption capacity. That passivation indeed
brings at least part of the metal back to normal is also apparent
from the O_2 chemisorption results and from temperature programmed
reduction (TPR) measurements. Figure 3 illustrates that even in the
SMSI state the metal quickly adsorbs O_2 at room temperature,
while a subsequent TPR experiment shows a reduction peak at − 50°C
([37]), which is attributed to the reduction of the outer layer of
the rhodium oxide on rhodium metal particles. The TPR profile also
shows a negative consumption of H_2 between 50 and about 250°C.
This must be due to H_2 desorption from the rhodium, proving that
metallic rhodium was present and attainable for H_2.
    Furthermore the fact that Ti^{3+} ions are quickly oxidized by
O_2 at room temperature, but that the SMSI state cannot be completely broken under the same conditions, suggests that the charge
transfer model for the explanation of SMSI (3, 13) is not very likely. The covering model ([5], [7], [14-19]), on the other hand, would not
be in contradiction to the results. Because, as suggested by
Mériaudeau et al. ([7]), the reduced TiO_x species which have been
formed and migrated over the metal surface during high temperature
reduction may very well be two dimensional in shape. During
reoxidation they will transform into three–dimensional TiO_2
particles on the metal surface. Thus even during room temperature
admission of O_2 part of the metal surface will be uncovered ([7]).

One may further presume that large metal particles will have a
thinner TiO_x layer and will be more easily uncovered than small
metal particles. During further O_2 admission at elevated temperatures the TiO_2 particles will migrate from the metal surface to
the support and will fully uncover the metal surface (cf. Figure 2).
  We want to end with a remark on the consequences of the
existence of a fast and slow H_2 adsorption on metal catalysts. The
existence of H_2 spillover to the support forces the scientist to
carefully design his experiments if he wants to extract only the
H_2 adsorption on the metal from his data, in order to determine
metal dispersion and metal particle size. In the past the proven or
suspected presence of spillover led to the method of measuring a
H_2 adsorption isotherm and extrapolating its high pressure part
back to zero pressure (24, 38). In another method a second isotherm
was measured after intermediate evacuation and the resulting,
so-called "reversible", H_2 adsorption was attributed to spillover.
The difference between first and second isotherm then was supposed
to be equal to the H_2 chemisorbed on the metal and was called
"irreversible" H_2 adsorption (39, 40). Even disregarding the fact
that there never exists such a thing as "irreversible adsorption",
the problem with both methods is that they give results which are
very dependent on the experimental conditions. Just because of the
fact that H_2 adsorption on the metal is fast and reversible, the
amount of H_2 adsorbed will depend on the final H_2 pressure and,
in the second method, on the pressure reached at the catalyst during
evacuation (41). For that reason H_2 adsorption measurements performed around 100 kPa always give higher results than measurements
performed around 10 kPa or even 1 kPa. Of course one has to take
into account also the possibility that at higher pressure more than
one hydrogen atom will be adsorbed per surface metal atom. Evidence
that the surface M:H stoichiometry may exceed one for some metals
has been growing lately (39, 40, 42, 43).
  All this leads to the conclusion that our knowledge of fractional coverage of hydrogen on a metal surface, of H:M stoichiometry
and of hydrogen spillover in supported metal catalysts is still limited and that careful studies to disentangle these three factors
should be encouraged. In this respect we feel that our result of a
slow H_2 spillover onto TiO_2 strongly suggests that a better method to separate H_2 adsorption on the metal from that on the
support would be to extrapolate H_2 adsorption measurments to zero
time, instead of to zero pressure.

Acknowledgment

This study was supported by the Netherlands Foundation for Chemical
Research (SON) with financial aid from the Netherlands Organisation
for the Advancement of Pure Research (ZWO).

Literature Cited

1. Tauster, S.J.; Fung, S.C.; Garten, R.L. J. Amer. Chem. Soc.
   1978, 100, 170.
2. Tauster, S.J.; Fung, S.C. J. Catal. 1978, 55, 29.
3. Tauster, S.J.; Fung, S.C;, Baker, R.T.K.; Horsley, J.A.
   Science 1981, 211, 1121.

4. Ko, E.I.; Garten, R.L. J. Catal. 1981, 68, 233.
5. Resasco, D.E.; Haller, G.L. J. Catal. 1983, 82, 279.
6. Mériaudeau, P.; Ellestad, O.H.; Dufaux, M.; Naccache, C.
    J. Catal. 1982, 75, 243.
7. Mériaudeau, P.; Dutel, J.F.; Dufaux, M.; Naccache, C.
    Stud. Surf. Sci. Catal. 1982, 11, 95.
8. Wang, S-Y.; Moon, S.H.; Vannice, M.A. J. Catal. 1981, 71, 167.
9. Vannice, M.A. J. Catal. 1982, 74, 199.
10. Burch, R.; Flambard, A.R. J. Catal. 1982, 78, 389.
11. Baker, R.T.K.; Prestridge, E.B.; Garten, R.L. J. Catal. 1979,
    59, 293.
12. Bardi, U.; Somorjai, J.A.; Ross, P.N. J. Catal. 1984, 85, 272.
13. Horsley, J.A. J. Amer. Chem. Soc. 1979, 101, 2870.
14. Santos, J.; Phillips, J.; Dumesic, J.A. J. Catal. 1983,
    81, 147.
15. Simoens, A.J.; Baker, R.T.K.; Dwyer, D.J.; Lund, C.R.F.;
    Madon, R.J. J. Catal. 1984, 86, 359.
16. Cairns, J.A.; Baglin, J.E.E.; Clark, G.J.; Ziegler, J.F.
    J. Catal. 1983, 83, 301.
17. Powell, B.R.; Whittington, S.E. J. Catal. 1983, 81, 382.
18. Sadeghi, H.R.; Henrich, V.E. J. Catal. 1984, 87, 279.
19. Chung, Y-W.; Xiong, G.; Kao, C-C. J. Catal. 1984, 85, 237.
20. Baker, R.T.K.; Prestridge, E.B.; Garten, R.L. J. Catal. 1979,
    56, 390.
21. Huizinga, T.; Prins, R. J. Phys. Chem. 1981, 85, 2156.
22. Apple, T.M.; Gajardo, P.; Dybowski, C. J. Catal. 1981,
    68, 103.
23. Conesa, J.C.; Soria, J. J. Phys. Chem. 1982, 86, 1392.
24. Benson, J.E.; Boudart, M. J. Catal. 1965, 4, 704.
25. Dowden, D.A. In "Specialists Periodical Reports–Catalysis";
    Kemball, C; Dowden, D.A., Eds.; The Chemical Society: London,
    1980; Vol. III, p. 136.
26. Bond, G.C. Stud. Surf. Sci. Catal. 1983, 17, 1.
27. Huizinga, T. Ph.D. Thesis, Eindhoven Univ. Technology, 1983.
28. Keren, E.; Soffer, A. J. Catal. 1977, 50, 43.
29. Duprez, D.; Miloudi, A. Stud. Surf. Sci. Catal. 1983, 17, 163.
30. Woning, J.; van Santen, R.A. Chem. Phys. Lett. 1983, 101, 541.
31. Iyengar, R.D.; Codell, M.; Gisser, H.; Weissberg, J.
    Z. Phys. Chem. N.F. 1974, 89, 324.
32. Jiang, X-Z.; Hayden, T.F.; Dumesic, J.A. J. Catal. 1983,
    83, 168.
33. Vis, J.C.; van 't Blik, H.F.J.; Huizinga, T.; Prins, R.
    J. Mol. Catal. 1984, 25, 367.
34. Vis, J.C.; van 't Blik, H.F.J.; Huizinga, T.; van Grondelle,
    J.; Prins, R. J. Catal. 1985, 95, 000.
35. Fromhold, A.T. In "Theory of metal oxidation I, series
    Defects in crystalline solids"; Amelinckx, S.; Gevers, R.;
    Nihoul J., Eds.; North Holland Publ. Comp.: Amsterdam, 1976;
    Vol. IX, p. 289.
36. Hauffe, K. In "Treatise on solid state chemistry"; Hannay,
    N.B., Ed.; Plenum Press: New York, 1976; Vol. IV, p. 389.
37. van 't Blik, H.F.J.; Vriens, P.H.A.; Prins, R. to be published.
38. Wilson, G.R.; Hall, W.K. J. Catal. 1970, 17, 190.
39. Sinfelt, J.H.; Via, G.H. J. Catal. 1979, 56, 1.

40. McVicker, G.B.; Collins, P.J.; Ziemiak, J.J. J. Catal. 1982, 74, 156.
41. Crucq, A.; Degols, L.; Lienard, G.; Frennet, A. Acta Chim. Acad. Sci. Hung. 1982, 111.
42. Wanke, S.E.; Dougharty, N.A. J. Catal. 1972, 24, 367.
43. van 't Blik, H.F.J.; van Zon, J.B.A.D.; Huizinga, T.; Vis, J.C.; Koningsberger, D.C.; Prins, R. J. Am. Chem. Soc. 1985, 107, 3139.

RECEIVED October 17, 1985

8

CO, O$_2$, and H$_2$ Heats of Adsorption on Supported Pd

M. Albert Vannice and Pen Chou

Department of Chemical Engineering, The Pennsylvania State University, University Park, PA 16802

Palladium dispersed on SiO$_2$, η-Al$_2$O$_3$, SiO$_2$-Al$_2$O$_3$, and TiO$_2$, was characterized by chemisorption measurements to determine H$_2$, O$_2$, and CO uptakes. Integral, isothermal heats of adsorption were then obtained using a modified differential scanning calorimeter. Values of $\Delta H_{(ad)}$ were typically near 55 ± 10 kcal/ mole O$_2$ and 25 ± 5 kcal/mole CO. Values for H$_2$ appeared to be around 20 kcal/mole H$_2$ after a correction was made for a baseline perturbation caused by the difference in thermal conductivities between H$_2$ and Ar. The $\Delta H_{(ad)}$ values on Pd/TiO$_2$ reduced at 773K were not markedly different from those on the "typical" Pd catalysts, and no significant support effect on heats of adsorption was found for Pd. For all three gases, heats of adsorption increased as Pd crystallite size decreased from 6 nm to below 2 nm. Pd hydride formation was routinely determined by chemisorption and calorimetric measurements and heats of formation of the β-phase hydride were consistently near 10 kcal/mole H$_2$ absorbed, independent of support and crystallite size, which is in good agreement with literature values for bulk palladium.

Since the report by Tauster et al. that high temperature reduction (HTR) near 773K markedly decreased the H$_2$ and CO chemisorption capacity of Group VIII metals dispersed on TiO$_2$ (1), many studies have been devoted to this phenomenon. One popular explanation is that bond strengths between the adsorbate and the metal are decreased as a consequence of an electronic effect caused by some form of electron transfer between the metal and the support (2-4). Another, which does not require a decrease in heats of adsorption, is that chemisorption is decreased because of physical blockage of metal surface sites created by the migration of a TiO$_x$ species onto the metal surface (5-7). In this study we wanted to conduct direct calorimetric measurements of CO and H$_2$ heats of adsorption on Pd dispersed on TiO$_2$ along with those obtained for "typical" Pd cata-

0097-6156/86/0298-0076$06.00/0

lysts utilizing SiO_2, Al_2O_3 and $SiO_2-Al_2O_3$ as supports. A comparison of these results should help determine whether a decreased heat of adsorption of CO and H_2 is the primary cause for lower chemisorption uptakes. It was also of interest to measure these isothermal, integral heats of adsorption because these values describe surfaces which are equilibrated at high gas pressures and, therefore, relate to surfaces which can exist under reaction conditions. Another reason for this investigation was the large variation in methanation turnover frequency which has been found for this family of Pd catalysts (8), and the correlation which has been observed between heats of adsorption and activity (9). It was therefore of interest to determine whether any relationship existed between these two parameters for this family of Pd catalysts.

Experimental

Catalyst Preparation. A number of the catalysts used in this study were prepared from $PdCl_2$ (Ventron Corp.) and $\eta-Al_2O_3$ (Exxon Research & Engineering Co.), $SiO_2-Al_2O_3$ (Davison Grade 979, 13% alumina), TiO_2 (P-25 from Degussa Co., 80% anatase and 20% rutile), and SiO_2 (Davidson Grade 57) using an impregnation technique (8). The reported surface areas for each support are 245, 400, 50 and 220 m^2 g^{-1}, respectively. To achieve maximum dispersion, an ion exchange technique with $Pd(NH_3)_4(NO_3)_2 \cdot 2$ H_2O was also used for several samples (10). The final Pd weight loadings of the catalysts were determined by both neutron activation analysis and atomic emission spectroscopy.

Chemisorption Measurements

Apparatus. The uptake measurements were made in a high vacuum system consisting of a glass manifold connected to an Edwards Model EO2 oil-diffusion pump backed by a mechanical pump with liquid nitrogen traps at the inlet of each pump. An ultimate dynamic vacuum near 4×10^{-7} torr (1 torr = 133 Pa) was obtainable, as measured by a Granville Phillips Model 260-002 ionization gauge. Isotherm pressures and temperatures were measured by a Texas Instruments Model 145 Precision Gauge and a Doric digital trendicator, respectively. A more detailed description of the gases, their purification, and the adsorption system is given elsewhere (11).
Procedures and Pretreatments. The "typical" Pd catalysts ($Pd/\eta-Al_2O_3$, Pd/SiO_2 and $Pd/SiO_2-Al_2O_3$) were given a pretreatment consisting of a 1 h reduction in flowing H_2 or a mixture of 20% H_2 + 80% He at 448K, 573K or 673K. Previous work had indicated that 448K was sufficient for reduction (1), but we found that better results were sometimes obtained at the higher temperatures (12). The Pd/TiO_2 catalysts were given either a low temperature reduction (LTR) at 448K or a high temperature reduction (HTR) at 773K in a flowing gas mixture of 20% H_2 and 80% He following the procedure of Tauster et al. (1). The Pd/TiO_2 (LTR) sample was treated in 20% O_2 at 573K for 1 h after each CO heat of adsorption measurement, prior to another (LTR) step, to stabilize the sample and to facilitate complete removal of the CO. After these pretreatments, adsorption was rapid as expected for nonactivated adsorption, and the pressure

typically stabilized within 15 min for the initial point and within
2–3 min for the succeeding uptakes at higher pressure.

Calorimetric Measurements

Isothermal energy changes were measured using a modified Perkin-
Elmer DSC-2C differential scanning calorimeter with an Intracooler
II (Model 319-0207) which allows subambient runs down to 200K. The
gas handling system preceding the calorimeter controlled the flows
of argon, helium, hydrogen and carbon monoxide and provided switching capability between gas streams, as shown in Figure 1. Ultra
high purity argon (99.999% from MG Scientific Gases) was further
purified by passing it first through a drying tube containing 5A
molecule sieve (Supelco Inc.), and then through an Oxytrap (Alltech
Associates) before use as a purge gas in the calorimeter. The
helium (99.9999% from MG Scientific) was purified in a similar
fashion. An Elhygen Mark V Hydrogen Generator produced ultra-pure
hydrogen (less than 10 ppb total impurities) by electrolytically
dissociating deionized water and then diffusing the hydrogen
through a thin palladium membrane. The carbon monoxide (99.99%
from Matheson) was passed through a molecular sieve trap held at
383K to remove any metal carbonyls. The gases were regulated by
Tylan mass flow controllers (Model FC260), and a digital Tylan
RO20A readout box provided control and monitoring for the four mass
flow controllers. Two of these controllers were capable of measuring between 0 and 50 cm^3 min^{-1} argon flow, the third one was
designed for either hydrogen or carbon monoxide flow control
between 0 and 10 cm^3 min^{-1}, depending upon which valve was opened,
and the fourth regulated He flows up to 10 cm^3 min^{-1}.

To enhance sensitivity and accuracy by minimizing baseline
perturbations after switching from the purge gas to the gas stream
containing the adsorbate, various evolutionary modifications were
made to the gas handling system and the calorimeter itself, and the
final flow design is shown in Figure 1. The changes to the DSC
included: 1) perforation of the platinum sample holder covers to
enhance gas mixing; 2) removal of the flow splitter inside the DSC
and insertion of a needle valve in each line (to the sample side
and to the reference side); 3) regulation of flows by mass flow
controllers; 4) utilization of adjustable He/Ar ratios in the purge
stream; and 5) enclosure of the entire aluminum block, the cover,
and the draft shield under a blanket of flowing N_2 (99% from Linde)
to eliminate the possibility of oxygen (air) diffusing through any
tiny leaks to the sample.

To provide optimum performance, the following procedure was
used. Needle valve 5 controlling gas flow to the atmosphere was
adjusted so that equal pressure drops were attained through both
loops of the switching valve. This eliminated a perturbation
during switching due to a pressure differential. Adjustment of the
two needle valves in lines 9 and 10 balanced the flows through the
sample and reference cavities at a constant overall flow rate. The
purge gas to the DSC was comprised of a constant Ar flow of ~ 36
cm^3/min which bypassed the switching valve plus an additional component which passed through the switching valve. For the experiments
with H_2, switching occurred from 8 cm^3 min^{-1} He to 4 cm^3 min^{-1} H_2

back to the He. These He flow rates (STP) minimized baseline
perturbation and offset due to differences between the thermal
conductivity of the mixture and pure Ar, a problem which was more
severe with the H_2 mixtures. After these modifications, very
reproducible results with minimal baseline correction were
obtained, as indicated in Figure 2 by the baseline traces after
purging both the sample and the reference sides and readsorbing the
gas.

   The signal output from the DSC and its time integral were
monitored on a 2-pen integrating recorder (Linear Instruments Model
252A). The energy calibration was conducted using various weights
of indium and different ordinate (mcal/s) ranges, chart speeds, and
heating rates to determine a calibration constant of K = 0.0442 ±
.0002 s/count for the recorder integrator. As a check, another K
value was determined based on the weight of the chart paper under
the curve. Both provided excellent correlations, so the former was
used in most cases. Eight ordinate sensitivities between 0.1 and
20 mcals^{-1} could be used, but with the sample sizes used here (0.03
- 0.15 g) and the typical gas uptakes that occurred, most runs
could be easily recorded on the 5 or 10 mcal s^{-1} full-scale range.
After admittance of H_2, O_2 or CO, the baseline was corrected by
subtracting by switching back to the purge gas for 1 h, reintroduc-
ing the adsorbate mixture, and obtaining the baseline trace after
weakly absorbed gas had been removed. For the runs with hydrogen,
this involved the determination of the energy associated with β-Pd
hydride formation, as shown in Figure 2(a).

   The samples placed in the DSC were taken from a larger sample
of that particular catalyst, which had been previously pretreated
and characterized in the chemisorption system then stored in a
desiccator. Each sample was then given another pretreatment in the
DSC in pure H_2 or a 20% H_2/80% Ar mixture, flowing at 40 cm^3 min^{-1}
and was heated at 40K min^{-1} to the desired temperature. Because of
the greater thermal conductivity of H_2, significant deviations
occurred between the actual cavity temperature and the temperature
indicated on the DSC when pure H_2 was used. For example, with a
flow of pure H_2, a maximum temperature of 713K was achieved rather
than the indicated 773K. A calibration between ΔT and the H_2 mole
fraction allowed the actual temperature to be obtained.

   Finally, to check for possible contamination effects, the
purge time at high temperature in the DSC system was varied from 20
min to 660 min as a N_2 blanket was kept around the cavity. Small
apparent increases in $\Delta H_{(ad)}$ for H_2 were observed as purge time
increased up to 60 min, which were attributed to an increase in
desorption from the surface, but additional purging to 180 min
produced no change, within experimental error, and a 660 min purge
resulted in a 5% increase in the energy change. Consequently, 60
min was chosen as an optimum time for complete removal of hydrogen
from the surface after reduction.

Results

A typical set of DSC traces is illustrated in Figure 2. The difference in areas under the trace obtained during adsorption and the
baseline trace represents the energy change associated with the

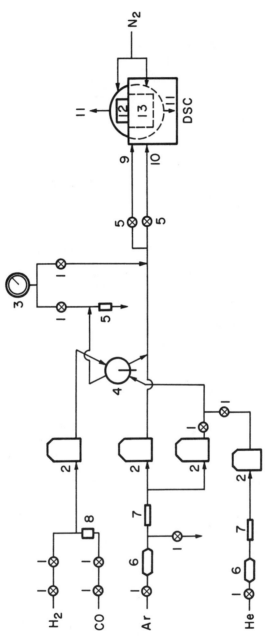

Figure 1. Differential Scanning Calorimetry System: 1--Hoke valves, 2--Mass flow controllers, 3--Pressure gauge, 4--Four-way switching valve, 5--Needle valves, 6--Supelco Drying tubes, 7--Oxytraps, 8--Molecular sieve trap, 9--Line to sample cavity, 10--Line to reference cavity, 11--N_2 purge streams through plastic shrouds, 12--Draft shield, 13--Aluminum block.

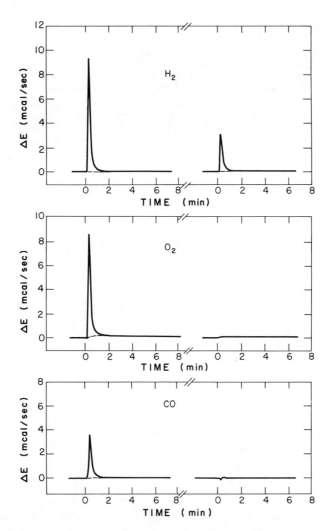

Figure 2. Energy Change During Adsorption at 300K on 1.95% Pd/SiO$_2$–Al$_2$O$_3$ (T$_R$ = 673K)

chemisorption process as the clean Pd surface was equilibrated with
the adsorbate at 75 torr (0.1 atm). In all cases the adsorption
process was essentially complete in less than 2 minutes, in agreement with our chemisorption experiments which showed that chemisorption on the Pd surface was very rapid.

A set of adsorption isotherms is shown in Figure 3. The
difference in uptakes at 75 torr was used to determine the amount
of gas irreversibly adsorbed on the Pd surface; however, both
isotherms are quite parallel over a wide pressure range. The
isotherm in Figure 3a clearly shows the onset of β-phase Pd hydride
formation at pressures above 10 torr but the amount of chemisorbed
hydrogen on either the β-phase or α-phase hydride remains constant.
The uptake from the lower isotherm, corrected for physical adsorption on the support, represents the extent of bulk hydride formation.

A set of isothermal, integral $\Delta H_{(ad)}$ values in kcal/mole is
listed in Table I for a series of Pd catalysts with an average Pd
crystallite size near 3 ± 1 nm, as measured from H_2 chemisorption.
More complete representations of measured $\Delta H_{(ad)}$ values for CO, O_2,
and H_2 are shown in Figures 4-6, respectively.

Table I. Integral Heats of Adsorption on Supported
Palladium (300K)

Catalyst	T_{Red} (°K)	Dia- meter (nm)	$H_{(ad)}$ (a) H_2	O_2	CO
2.1% Pd/SiO$_2$	573	1.7	32	70	29
0.48% Pd/SiO$_2$	573	2.4	35	60	30
1.95% Pd/SiO$_2$–Al$_2$O$_3$	448	3.2	32	55	22
1.95% Pd/SiO$_2$–Al$_2$O$_3$	673	4.0	33	50	22
1.80% Pd/η–Al$_2$O$_3$	673	3.4	31	61	20
2.03% Pd/TiO$_2$	448	3.7	33	71	--
2.03% Pd/TiO$_2$ (O$_2$ @573K)	448	3.6	32	62	24
2.03% Pd/TiO$_2$	773	(3.7)	26	--	17

(a) Apparent values prior to any correction

Finally, a series of values for the apparent enthalpy of formation of β-phase Pd hydride, along with the bulk Pd hydride ratios,
is listed in Table II. The bulk ratios were obtained by dividing
the hydrogen uptake at 75 torr from the second isotherm by the
number of bulk Pd atoms, i.e., the total number of Pd atoms minus
the surface Pd atoms, as determined by chemisorbed hydrogen.

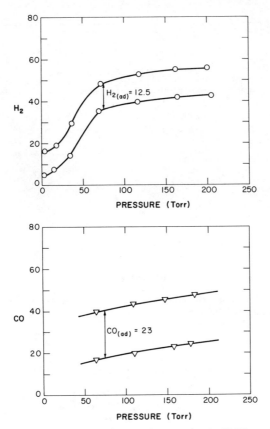

Figure 3.   Gas Uptake (μmole/g cat.) at 300K on 1.95% Pd/SiO$_2$-Al$_2$O$_3$ (T$_R$ = 673K).

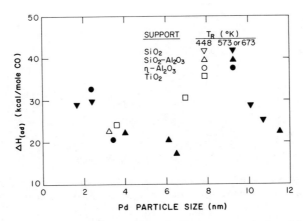

Figure 4.   Heat of adsorption of CO on supported palladium (300K).

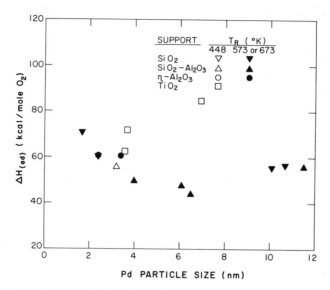

Figure 5.   Heat of adsorption of oxygen on supported palladium (300K).

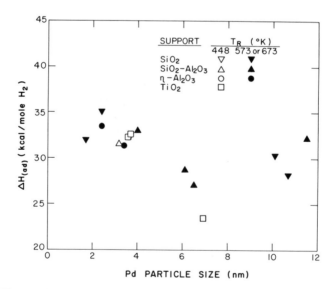

Figure 6.   Heat of adsorption of hydrogen on supported palladium (300K). Apparent values prior to correction.

Table II. Enthalpy of Formation of Palladium Hydride

Catalyst	T_{Red} (°C)	Diameter (nm)	$\dfrac{H(ab)}{Pd_{bulk}}$	$\dfrac{\Delta H_f}{\begin{array}{c}\text{kcal}\\ \text{mole } H_2\end{array}}$ (a)
2.10% Pd/SiO$_2$	400	1.7	0.83	25.3
0.48% Pd/SiO$_2$	300	2.4	0.87	21.5
1.95% Pd/SiO$_2$-Al$_2$O$_3$	175	3.2	0.63	22.4
"	400	6.5	0.43	20.6
1.80% Pd/η-Al$_2$O$_3$	400	3.4	0.57	20.3
2.03% Pd/TiO$_2$	175	3.6	0.65	17.9
Pd Powder	200	1μ	0.69	21.3 (Abs.)
				20.6 (Des.)

(a)Apparent uncorrected values.

Discussion

Initial heats of adsorption, obtained under UHV conditions on Pd films and single crystals, have varied from 32-40 kcal/mole CO, 55-80 kcal/mole O$_2$, and 21-26 kcal/mole H$_2$ (13). Because $\Delta H_{(ad)}$ values are known to decrease with increasing surface coverage, especially for CO and O$_2$, the integral heats of adsorption measured here are expected to be lower than the initial values. This behavior is observed for CO and O$_2$ as all values are less than the reported maximum values. These results supplement those recently reported (14). The one high value for the Pd/TiO$_2$ sample in Figure 5 is undoubtedly due to some adsorption on a partially reduced titania surface. A metal like Pd can catalyze oxygen removal from the titania lattice, hence the pure TiO$_2$ used as a blank does not give completely satisfactory correction for O$_2$ adsorption on the titania surface. Only one previous study has reported $\Delta H_{(ad)}$ values for supported Pd catalysts (15), and the values of 26.3 kcal/mole H$_2$ and 40-80 kcal/mole O$_2$, determined by calorimetry for a Pd/MgO system, are in good agreement with those in our investigation. Somewhat surprisingly, the H$_2$ heats of adsorption were approximately 10 kcal/mole higher than anticipated, based on previous values. The reason for this is still not clear; however, we believe that this may be due to an artifact. Experiments now underway indicate that this may be caused by the initial removal of H$_2$, which has a very high thermal conductivity, from the Ar carrier stream, which has a much lower thermal conductivity. The error appears to be proportional to the H$_2$ taken up by the catalyst and seems to be about 10 kcal/mole H$_2$, which lowers the apparent $\Delta H_{(ad)}$

to the anticipated region. This is not a problem for CO and O_2 because of the similarity of thermal conductivities with that of Ar.

When plotted versus crystallite size, independently of the support used, a similar crystallite size effect appears to exist for all three adsorbates–– heats of adsorption increase somewhat as crystallite size decreases from 6 nm to below 2 nm. All Pd crystallite sizes were based on hydrogen chemisorption and the assumption of 1.2×10^{19} $Pd_s m_2^{-1}$ (8). At this time we believe there is a possibility that the values reported for the most poorly dispersed samples––particle sizes of 10 nm and higher––may actually represent a bimodal crystallite distribution because of their preparation by sintering in H_2. Consequently, the observed $\Delta H_{(ad)}$ values would be higher than expected for large supported Pd crystallites. Characterization by TEM is now underway to better characterize these catalysts. Further study is necessary to determine the reasons for this apparent crystallite size effect on heats of adsorption, especially since the trend is opposite to that found in supported Pt catalysts (16,17).

One of the most interesting aspects of this study is the small influence of the support on $\Delta H_{(ad)}$ values, as indicated in Table I. Here a comparison is made at approximately constant crystallite size to eliminate variations due to that parameter. For the "typical" catalysts, essentially no influence is observed for H_2 and O_2 adsorption. The small variation which exists for CO adsorption, with the highest $\Delta H_{(ad)}$ values occurring for Pd/SiO_2 catalysts, is consistent with behavior observed for Pt (17). However, in contrast to the Pt catalysts, an HTR treatment for Pd/TiO_2 decreases $\Delta H_{(ad)}$ for both H_2 and CO by only about 5–7 kcal/mole; therefore, H–Pd and CO–Pd bond strengths remain high on this "SMSI" catalyst and the greatly diminished chemisorption capacity after the HTR step (1,8,14) cannot be attributed to a weak adsorbate–metal interaction. This leaves physical blockage of the Pd surface by migrating TiO_x species as the principal reasons for decreased chemisorption, a conclusion consistent with recent studies (5–7,18).

Because of the insensitivity of H_2 heats of adsorption to the support, there is no correlation between CH_4 turnover frequencies determined earlier (8) and $\Delta H_{(ad)}$ values for H_2. Also, the random variation among $\Delta H_{(ad)}$ values for CO also precludes any clear correlation between CH_4 turnover frequencies and this parameter. Consequently, we propose that the enhanced activity is due to special active sites created at the $Pd–TiO_x$ interface, as proposed earlier for Pt (19).

Finally, the enthalpies of formation for β–Pd hydride are very intriguing. They are not especially sensitive to either crystallite size or to the support, and the apparent values associated with typical H/Pd_b ratios near 0.6–0.7 are repeatedly near 20 kcal/mole H_2; consequently, the corrected values are close to 10 kcal/mole H_2, in agreement with the literature (20–22). These values are also in excellent agreement with those we obtained using ultrapure Pd powder (Puratronic grade, 99.999% from Johnson-Matthey), for which the exotherm obtained during absorption and the endotherm measured during decomposition under pure flowing Ar were in excellent agreement (Table II). The use of Pd powder to measure hydride formation provides a valuable internal standard to check the calibration of our instrument.

Summary

We have found that a modified DSC-2C calorimeter system is very applicable to supported metal catalysts, and reproducible heats of adsorption can be rapidly determined for many gases. Our values for CO and O_2 are in very good agreement with values measured in UHV systems, being consistently lower than these initial values obtained at low coverages; however, the apparent heats of adsorption for H_2 were always higher than anticipated by about 10 kcal/mole H_2, and a correction of this size was required for H_2 when Ar was used as a carrier gas. Heats of adsorption for all three gases increased as crystallite size decreased from 6 nm to below 2 nm. No major support effect on $\Delta H_{(ad)}$ values was observed for Pd, and no correlation was found between CH_4 turnover frequency and H_2 or CO heats of adsorption, in contradistinction with Pt (16,17). This implies that generalizations cannot be made for all Group VIII metals on TiO_2, even for the noble metals, and each metal/TiO_2 system must be individually studied. These results strongly support the explanation that the decrease in chemisorption capacity on TiO_2-supported Pd is primarily due to site blockage by TiO_x species migrating onto the Pd surface.

Literature Cited

1. Tauster, S. J.; Fung, S. C.; Garten, R. L. J. Am. Chem. Soc. 1978, 100, 170.
2. Tauster, S. J.; Fung, S. C.; Baker, R. T. K.; Horsley, J. A. Science. 1981, 211, 1121.
3. Santos, J.; Phillips, J.; Dumesic, J. A. J. Catal. 1983, 81, 147.
4. Sachtler, W. M. H. Proc. 8th Int. Cong. Catal., Dechema, Frankfurt am Main, 1984, I-151.
5. a) Meriaudeau, P.; Dutel, J.; Dufaux, M.; Naccache, C. In "Studies in Surface Science and Catalysis"; Imelick, B., et al., Eds.; Elsevier, New York, 1982; Vol. 11.
    b) Meriaudeau, P.; Ellestad, O. H.; Dufaux, N.; Naccache, C. J. Catal. 1982, 75, 243.
6. Resasco, D. E.; Haller, G. L. J. Catal. 1983, 82, 279.
7. Jiang, X.-Z.; Hayden, T. F.; Dumesic, J. A. J. Catal. 1983, 83, 168.
8. Wang, S-Y.; Moon, S. H.; Vannice, M. A. J. Catal. 1981, 71, 167.
9. Vannice, M. A. J. Catal. 1977, 50, 228.
10. Benesi, H. A.; Curtis, R. M.; Studer, H. P. J. Catal. 1968, 10, 328.
11. Palmer, M. B.; Vannice, M. A. J. Chem. Tech. Biotech. 1980, 30, 205.
12. Sudhakar, C.; Vannice, M. A. Appl. Catal. 1985, 14, 47.
13. Toyashima, I.; Somorjai, G. A. Catal. Rev.-Sci. Eng. 1979, 19, 105.

14. Vannice, M. A.; Chou, P. JCS Chem. Comm. 1984, 1590.
15. Zakumbaera, G. D.; Zakarina, N. A.; Naidin, V. A.; Dostiyarov, A. M.; Toktabaeya, N. F.; Litvyakova, E. N. Kinet. Katal. 1983, 24, 379.
16. Vannice, M. A.; Hasselbring, L. C.; Sen, B. J. Catal. In press.
17. Vannice, M. A.; Hasselbring, L. C.; Sen, B. J. Catal. Submitted for publication.
18. Baker, R. T. K.; Prestridge, E. B.; McVicker, G. B. J. Catal. 1984, 89, 422.
19. Vannice, M. A.; Sudhakar C. J. Phys. Chem. 1984, 88, 2429.
20. Nace, D. M.; Aston, J. G. J. Am. Chem. Soc. 1957, 78, 3619.
21. Picard, C.; Kleppa, O. J.; Boureau G. J. Chem. Phys. 1978, 69, 5549.
22. Kuji, T.; Oates, W. A.; Bowerman, B. S.; Flanagan, T. B. J. Phys. F. 1983, 13, 1785.

RECEIVED September 12, 1985

Equilibrium and Kinetic Aspects of Strong Metal–Support Interactions in Pt–TiO$_2$ and Cobalt-Doped Cu–ZnO–Al$_2$O$_3$ Catalysts

M. S. Spencer

Agricultural Division, ICI PLC, Billingham, Cleveland TS23 1LB, England

Equilibrium and kinetic calculations, based on bulk properties, support the view that strong metal–support interactions in Pt/TiO$_2$ catalysts arise from the decoration of platinum crystallites with TiO$_x$ species. The surface concentration of titanium atoms in dilute Pt/Ti alloys is negligible <u>in vacuo</u> but traces of water, present under normal experimental conditions, are sufficient to oxidise titanium atoms, so causing extensive surface segregation of Ti as TiO$_x$. Rates of diffusion across and through the platinum, extrapolated from bulk data, are too slow under low-temperature reduction to give SMSI states in Pt/TiO$_2$ catalysts. Experimental results indicate that diffusion of TiO$_x$ across platinum is more important than diffusion of Ti through platinum in the formation of the SMSI state. The inhibition of chemisorption and methanol synthesis activity in Cu/ZnO/Al$_2$O$_3$ catalysts doped with small concentrations of cobalt is interpreted by a similar model of SMSI.

The inhibition of chemisorption and catalysis by Pt/TiO$_2$ catalysts after high-temperature reduction (HTR) is now widely attributed ([1-7]) to the local geometric and electronic effects of a monolayer (or less) of a partially-reduced titanium oxide, TiO$_x$, decorating the surface of the platinum crystallites. In principle this monolayer can be formed in two ways. If reduction conditions lead to the formation of Pt/Ti alloys, then the partial oxidation of titanium atoms in the platinum surface could give a surface titanium atom bonded both to platinum atoms and adsorbed oxygen, ie adsorbed TiO$_x$. Alternatively, diffusion of TiO$_x$ across the surface of the platinum crystallites from the partially-reduced TiO$_2$ support could give the same final state. For convenience of calculation of equilibrium properties this final state is assumed to be formed via Pt/Ti alloys. Bulk properties are used to deduce probable microscopic states, in both equilibrium and kinetic aspects, and these are compared with recent experimental results.

0097-6156/86/0298-0089$06.00/0
© 1986 American Chemical Society

The models are extended from Pt/TiO$_2$ to some other systems for which data are available. The inhibition of Cu/ZnO/Al$_2$O$_3$ methanol synthesis catalysts by low levels of Group VIII metals, especially cobalt, is shown to fit the same pattern of SMSI: the active metal surface is wetted by a reduced oxidic layer from the support.

Equilibrium States of Platinum Crystallites

Although reduction of TiO$_2$ to a stoichiometric lower valent titanium oxide, much less elemental titanium, is not possible under usual HTR conditions, the formation of Pt/Ti alloys is possible, extending as far as the formation of Pt$_8$Ti and Pt$_3$Ti, as recognised originally by Tauster et al (8). This is due mainly to the very strong Pt–Ti bonding in these alloys, usually attributed to strong interactions between the electrons in the d bands of both metals, which has been predicted theoretically (9) and shown experimentally (10). The consequences of intermetallic compound formation on chemisorption and catalysis depend strongly upon the composition of the surface layer, ie, on the exent to which surface segregation of either titanium or platinum may occur. The limited experimental evidence is considered later after an analysis based on bulk properties.

Simple criteria for surface segregation in alloys (relative melting points, enthalpies of sublimation, metal atom radii, surface free energies of the pure metals) all indicate that surface segregation of titanium should occur on Pt/Ti alloys in vacuo. However, this is inadequate because of the large departures from ideality in Pt/Ti alloys. Analysis (11) of a broken bond model of the system, especially with the use of data directly determined with Pt/Ti alloys, gives a more reliable result.

The free energy of segregation can be regarded (13) as the sum of two components, a strain energy from the mismatch of atom sizes and a surface free energy term due to the differences in atom–atom bond energies. Although the radius of a titanium atom in the pure metal is larger than the equivalent radius for platinum, the radius of a titanium atom in Pt$_8$Ti is less than that of the platinum atoms. The contribution from strain energy to the free energy of segregation is therefore negligible (12). A consideration of the nearest–neighbour bonds broken and formed in the exchange of platinum and titanium atoms between the surface layer and the bulk, for a very dilute solution of titanium in platinum, gives the main contribution to the free energy of segregation:

$$\Delta G_s = \Delta Z \ (G_{PtTi} - G_{PtPt}) \tag{1}$$

ΔZ is the difference in effective coordination number of a bulk site and a surface site, G_{PtTi} is the nearest–neighbour bond strength between Pt and Ti atoms (expressed as free energy, not enthalpy) and G_{PtPt}, that between Pt atoms. With values derived from pure platinum and from Pt$_8$Ti, the calculated values (13) of the segregation ratio (ie the ratio of surface and bulk concentration) vary between 2×10^{-7} for the (110) face to 10^{-4} for the (111) face at 800 K, and 5×10^{-12} to 2×10^{-7} for the same faces at 500 K.

Thus the surfaces of dilute Pt/Ti alloys in vacuo (or indeed in pure hydrogen) consist essentially of pure platinum. As the strength of the Pt-Ti bond is much greater than that of either Pt-Pt or Ti-Ti bonds, the most stable system, maximising the number of Pt–Ti bonds, is obtained when all the titanium atoms are in the bulk, surrounded by platinum atoms.

Experimental results bear out these theoretical predictions. Work with TiO_2 deposited on platinum foil (14, 15) indicates that under UHV conditions in the absence of oxygen little, if any, segregation of titanium occurs, in agreement with the theoretical predictions (11). In contrast, an AES study (16) of stoichiometric Pt_3Ti showed that the surface composition is close to the bulk composition, with consequent effects on H_2 and CO chemisorption. The theoretical analysis (11) of surface segregation in Pt_3Ti, on the same basis as the analysis of dilute solid solutions, predicts no surface segregation at exact stoichiometry. The broken bond model of surface segregation is also successful in the prediction of surface segregation in some other non-ideal platinum alloys (17).

The surface properties of dilute solid solutions of titanium in platinum should therefore closely resemble pure platinum, so alloy formation as such, if it occurs, can provide no ready explanation of SMSI effects.

The Effects of Water on Platinum/Titanium Crystallites

Traces of water, present under normal HTR conditions, may lead to surface oxidation of the Pt/Ti crystallites, to limits set by the H_2/H_2O ratio. The effect of surface oxidation on the surface segregation of titanium can be calculated (13) by the use of a modified broken-bond model in which any adsorbed oxygen is bonded primarily to a titanium atom in the platinum surface. (Ti-O bond energies are greater than Pt-O bond energies.) Although exact calculation is not possible because the strength of any O_a-Ti(Pt) bond is not known experimentally, some limits can be deduced which indicate qualitatively the surface state of the metal crystallites. If the free energy of an O_a-Ti(Pt) bond is assumed to be in the same range as those of Ti-O bonds in titanium oxides, then extensive surface oxidation and the consequent segregation of titanium at the surface (eg segregation ratios of $>10^3$) can occur with a few ppm only of water in a gas-phase of hydrogen. Thus the equilibrium state of platinum crystallites upon a titania support, produced under the usual HTR conditions, is that of decoration by a near monolayer of reduced titanium oxide, TiO_x, stabilised by strong Pt-Ti bonding. Under much more reducing conditions, eg in a UHV apparatus, little surface oxidation occurs and indeed any titanium oxide film, present on the platinum surface, would be reduced to form Pt/Ti alloys. This has now been shown experimentally with TiO_2 deposited on platinum foil (14, 15). Re-oxidation of the SMSI state, either by molecular oxygen or higher levels of water (eg from CO/H_2 reactions) leads to breaking of the Pt-Ti bonds and removal of the monolayer as TiO_2. It is not known whether the reformed TiO_2 stays as small crystallites on the platinum surface or returns to

the support phase.   This then allows the recovery of the normal
chemisorption properties of platinum upon low-temperature rereduction.

Encapsulation of Metal Crystallites

The TiO_x monolayer model differs from that of catalyst deactivation
by encapsulation from the support, which is usually irreversible.
Encapsulation has been known for many years, eg in Ni/SiO_2 catalysts
(18), and understood, for Ni/SiO_2 catalysts, as an initial "wetting"
of the metallic nickel surface by a nickel hydrosilicate film,
followed by accretion of silica to give a coating many layers thick.
Unreduced nickel hydrosilicate may also contribute to the
interaction between nickel and silica, depending on catalyst
preparation (19).   The driving force for this process is the
decrease in surface energy, first by the replacement of the nickel
metal surface by a siliceous surface and then by the decrease of
total silica surface area (as metal particles are normally smaller
than the support particles), analogous to Ostwald ripening.

   In comparison, the unexpected feature from HTR of Pt/TiO_2
catalysts is not the coverage of metal crystallites by oxide, but
the failure of the coverage to extend beyond a monolayer.   The
reduced oxide, TiO_x, cannot be present as a bulk phase because
normal HTR conditions are too oxidising (13) ie, the monolayer of
TiO_x requires Pt-Ti bonding for stability.   It seems possible that
difficulties of epitaxial fit may account for the failure of a
coating of TiO_2 to grow from the TiO_x monolayer of the SMSI state.
These difficulties could arise from either an incomplete coverage of
the platinum surface by the TiO_x layer or a mismatch of lattice
parameters between the TiO_x layer and TiO_2.   Titania forms glasses
or amorphous phases, which could cover a TiO_x/TiO_2 boundary, much
less readily than silica.

Kinetics of the Formation of the SMSI State

Equilibrium calculations for 500 K show that the SMSI state could
form under LTR as well as HTR conditions (13), so the failure to
achieve this in experiments with Pt/TiO_2 catalysts must be explained
in terms of the kinetics of the system.

   The mobility of the platinum surface itself is probably the
critical factor in the rate of diffusion of TiO_x across the platinum
surface.   The diffusing entity could be $OTi(Pt)_n$.   As Ti^{3+} can be
detected (20) in the TiO_2 support after LTR, the reduction step is
probably not limiting attainment of the SMSI state.   Nevertheless,
the ease of reduction of the support may be significant with
different supports, eg, the facile formation of the SMSI state in
Ni/Nb_2O_5 catalysts (7) may well be due to the reducibility of Nb_2O_5.
Resasco and Haller (3) found that deactivation of Rh/TiO_2 catalysts
after HTR depended on the square root of reduction time, as would be
expected if the rate of formation of the SMSI state is controlled by
a diffusion process rather than a reduction step.   Small metal
crystallites are also deactivated more readily than large
crystallites.

From work on the surface self–diffusion in metals, eg (21), mobility on a platinum surface would be expected to be significant above about 700 K (the Huttig temperature), so giving rapid surface diffusion of TiO_x across a platinum surface at 800 K (HTR) but not at 500 K (LTR). Other experimental results tend to confirm this model. After reduction at 773 K the most refractory of the platinum group metals, osmium (Huttig temperature about 1100 K), when supported on titania, still had about 50% of the chemisorption capacity generated by LTR (8). Ruthenium (Huttig temperature about 740 K) gives a marginal case. Variation of reduction temperature of a Ru/V_2O_3 catalyst between 523 and 873 K brought about only small changes in catalytic activity and selectivity (22). Bond and Yide (23) found that a reduction temperature of 893 K was needed initially to produce SMSI in ruthenium/titania catalysts. Subsequent oxidation of these catalysts gave better dispersions of ruthenium oxide over the titania, the reduction of which afforded more highly dispersed metal in smaller crystallites, which easily succumbed to SMSI in further high temperature reduction, eg at 758 K. In contrast, an SMSI state can be formed in some supported copper catalysts (see below) by reduction at 500 K, above the Huttig temperature for copper (about 450 K). However, some Cu/TiO_2 catalysts have been found (24) to reach a SMSI state after reduction at 773 K but not at 573 K. A failure to achieve sufficient reduction of the titania with the weaker hydrogenating metal may be the explanation. The effect of potassium in inducing an SMSI state in Pt/TiO_2 catalysts under LTR is probably due to increased mobility of TiO_2 or TiO_x on the platinum surface (25). A suboxide of titanium and silver have the same activation energy of migration over rhodium particles (26), indicating a common dependence on the surface mobility of the metal.

The experimental evidence therefore, favours surface diffusion of TiO_x as the main mechanism of the formation of the SMSI state under the normal HTR conditions. Nevertheless, a similar difference in rates for LTR and HTR can be calculated for the alternative mechanism and Cairns et al (6) have shown experimentally that TiO_x can be formed via Pt/Ti alloys at higher temperatures. Estimates (13) for the rate of diffusion of titanium metal through platinum are uncertain but indicate that significant diffusion in platinum crystallites would occur in experimental times only under HTR. However, the fast diffusion of a partially oxidised titanium species through bulk platinum has been observed experimentally (14, 15) and this appears to be faster than the diffusion of metallic titanium. Similar observations have been made on the Ni/TiO_2 system (27). These results are surprising for diffusion through metals but as polycrystalline foils were used in both studies the path of diffusion of TiO_x may have been along grain boundaries.

Validity of the Use of Bulk Properties

The application of bulk properties to predict the behaviour of microscopic systems can introduce errors both from the presence of

surfaces and from the very small particle size. Similar
difficulties arise in the use of macroscopic foil or single crystals
in attempts to simulate supported metal catalysts. However, it is
shown elsewhere (13) that likely errors do not affect the validity
of the models of the processes which could occur in Pt/TiO$_2$
catalysts.

SMSI Effects in Supported Copper Catalysts

Experimental and Results. The addition of substantial amounts of
cobalt to supported copper catalysts of type used for methanol
synthesis gives catalysts capable of producing higher alcohols as
well as methanol, although the systems display some complexities
(28).   Low levels of cobalt, eg 0.1–5% wt/wt, added to industrial
Cu/ZnO/Al$_2$O$_3$ LP methanol synthesis catalysts (either by
incorporation of cobalt nitrate in the co-precipitation or by
subsequent impregnation of cobalt nitrate or acetate solutions) lead
to a loss of methanol synthesis activity, without the formation of
methane or any other organic products.   Experimental conditions for
catalyst reduction and test of catalyst activity with CO/CO$_2$/H$_2$
mixtures have been described (30).   Reduction conditions
corresponded to LTR.   Lin and Pennella (29) have also observed the
deactivation of several copper catalysts by low levels of cobalt.
Similar effects are shown by the addition of some other Group VIII
metals, eg nickel.   As copper is soluble in metallic cobalt (about
8% at 500 K) the inhibition of methanation, etc is understandable
from the performance of nickel-copper alloy catalysts. However, as
cobalt is essentially insoluble in metallic copper (<0.1% at 500 K),
alloying of the copper crystallites cannot account for the loss of
methanol synthesis activity, for which copper metal surface is
required (30).

All techniques of bulk analysis used indicated that the
catalysts were unaffected by cobalt doping, eg X-ray diffraction
analysis showed copper metal and ZnO crystallites of the customary
sizes, with no evidence of cobalt-containing phases.   However,
marked effects of cobalt addition were evident when surface
techniques were applied.   Both hydrogen and carbon monoxide
chemisorption (measured at 293 K) fell to zero, in a similar way to
methanol synthesis activity, with increasing cobalt content (Figure
1).   ESCA and SIMS analysis of unreduced and reduced catalysts
showed that surface concentrations of cobalt were much higher than
the bulk concentration.   Titration with N$_2$O of a reduced catalyst
containing 5% Co, a technique used (30) to determine copper surface
areas, gave a negligible surface area.

Discussion. These results show clearly that although copper
crystallites are present in the reduced, cobalt-doped catalysts, the
copper surface is not available for catalysis or chemisorption.
Also, the added cobalt, although present in the surface layers in
significant concentration, is in an oxidised form and not metallic,
even though cobalt oxides should be reduced to cobalt metal under
catalyst activation conditions.   It is probable that the surface of
the copper crystallites was largely covered by an oxidic layer
containing cobalt and support (Al, Zn) oxides.

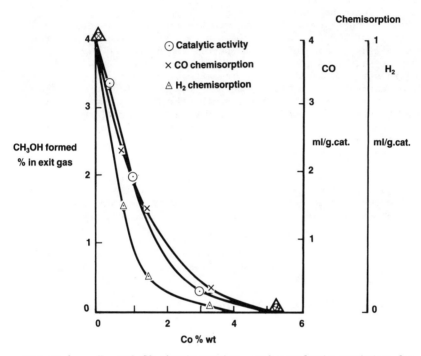

Figure 1.   H₂ and CO chemisorption, and catalytic activity for methanol synthesis by Cu/ZnO/AL₂O₃ catalysts with added cobalt.

In standard reduced $Cu/ZnO/Al_2O_3$ catalysts the copper surface
is not wetted by any of the oxidic components. The addition of
cobalt could give the formation of an oxidic Co/Zn/Al phase (eg a
"two-dimensional spinel") which, while not readily reduced to
metallic cobalt, is partially reducible. Chin and Hercules (31)
have described the formation of a cobalt "surface spinel", enhanced
by the presence of zinc, in Co/Al_2O_3 catalysts. This material, by
virtue of Co–Cu bonding, then wets the copper metal surface,
inhibiting chemisorption. Although the electronic interaction
between a Cu(100) face and a cobalt monolayer is weak (33), strong
Co–Cu bonding is not required to stabilise the oxidic layer on
copper crystallites (the equilibrium state of cobalt under the test
conditions is the metal). The resemblance to the SMSI state of
Pt/TiO_2 catalysts is obvious, but here the condition is produced by
LTR (so the inhibition is not reversible) and it is sufficiently
strong to prevent a synthesis gas reaction. Encapsulation of
supported nickel catalysts by a layer of nickel aluminate has been
described (34).

Conclusions

1   The equilibrium and surface segregation calculations support the
     view that the SMSI effects in Pt/TiO_2 catalysts are due to a
     monolayer (or less) of strongly-adsorbed titanium suboxide on
     the platinum surface.

2   The failure of Pt/TiO_2 catalysts to show SMSI effects after low
     temperature reduction is due to kinetic rather than equilibrium
     limitations.

3   Two ways of forming the TiO_x layer are possible:

     (a)   Formation of TiO_x at the Pt/TiO_2 interface, followed by
            transport across the Pt surface.

     (b)   Reduction of TiO_2 to give a Pt/Ti alloy, followed by
            diffusion, surface segration and oxidation of Ti.

     Both routes are sufficiently fast in HTR but both are too slow
     in LTR. Experimental evidence now shows that (a) is more
     important than (b), under the conditions typical of high
     temperature reduction.

4   The inhibition of chemisorption and methanol synthesis by cobalt
     addition to $Cu/ZnO/Al_2O_3$ catalysts occurs by a similar mechanism
     in which copper crystallites are covered by a cobalt-containing,
     oxidic monolayer.

Acknowledgments

G C Chinchen, P Snowden of ICI Agricultural Division, and R Marbrow,
J Myatt and M H Stacey of ICI Corporate Laboratory are thanked for
help with various aspects of the work with copper/cobalt catalysts.

Literature Cited

1   Meriaudeau, P.; Dutel, J. F.; Dufaux, M.; Naccache, C. In
    "Studies in Surface Science and Catalysis" Vol. 11, "Metal-
    Support and Metal-Additive Effects in Catalysis", Elsevier,
    Amsterdam, 1982; p. 95.

2   Santos, J.; Phillips, J.; Dumesic, J. A. J. Catal. 1983, 81,
    147.

3   Resasco, D. E.; Haller, G. L.   J. Catal. 1983, 82, 279; Appl.
    Catal. 1983, 8, 99.

4   Huizinga, T.; van't Blik, H. F. J.; Vis, J. C.; Prins, R.
    Surface Sci. 1983, 135, 580.

5   Belton, D. N.; Sun. Y. M.; White, J. M. J. Am. Chem. Soc. 1984,
    106, 3059.

6   Cairns, J. A.; Baglin, J. E. E.; Clark, G. J.; Ziegler, J. F.
    J. Catal. 1983, 83, 101.

7   Ko, E. I.; Hupp, J. M.; Wagner, N. J. J. Catal. 1984, 86, 315.

8   Tauster, S. J.; Fung, S. C.; Garten, R. L. J. Am. Chem. Soc.
    1978, 100, 170.

9   Brewer, L.; Wengert, P. R. Metall. Trans. 1973, 4, 83.

10 Meschter, P. J.; Worrell, W. L. Metall. Trans. 1976, 7A, 299.

11 Spencer, M. S. Surface Sci. 1984, 145, 145.

12 Abraham, F. F.; Tsai, N. H.; Pound, G. M. Surface Sci. 1979,
    83, 406.

13 Spencer, M. S. J. Catal. 1985, 93, 215.

14 Ko, C. S.; Gorte, R. J. J. Catal. 1984, 90, 59.

15 Demmin, R. A.; Ko, C. S.; Gorte, R. J. J. Phys. Chem. 1985, 89,
    1151.

16 Dardi, U.; Somorjai, G. A.; Ross, P. N. J. Catal. 1984, 85,
    272.

17 Spencer, M. S. Surface Sci. 1984, 145, 153.

18 Schuit, G. C. A.; van Reijen, L. L. Adv. Catal. 1958, 10, 243.

19 Blackmond, D. G.; Ko, E. I. Appl. Catal. 1984, 13, 49.

20 Huizinga, T.; Prins, R. J. Phys. Chem. 1981, 85, 2156.

21   Schrammen, P.; Holzl, J. Surface Sci. 1983, 130, 203.

22   Kikuchi, E.; Nomura, H.; Matsumoto, M.; Morita, Y. Appl. Catal.
     1983, 7, 1.

23   Bond, G. C.; Yide, Y. J. Chem. Soc., Faraday Trans. I. 1984,
     80, 3103.

24   Delk, F. S.; Vavere, A. J. Catal. 1984, 85, 380.

25   Spencer, M. S. J. Phys. Chem. 1984, 88, 1046.

26   Rouco, A. J.; Haller, G. L. J. Catal. 1981, 72, 246.

27   Takatani, S.; Chung, Y. W. J. Catal. 1984, 90, 75.

28   Courty, P.; Durand, D.; Freund, E.; Sugier, A. J. Mol. Catal.
     1982, 17, 241.

29   Lin, F. N.; Pennella, F. In "Catalytic conversion of synthesis
     gas and alcohols to chemicals"; R. G. Herman, Ed.; Plenum Press:
     New York, 1984; p. 53.

30   Chinchen, G. C.; Denny, P. J.; Parker, D. G.; Short, G. D.;
     Spencer, M. S.; Waugh, K. C.; Whan, D. A. PREPRINTS, Div. of
     Fuel Chem., Amer. Chem. Soc. Meeting, August 1984,
     Philadelphia, PA.

31   Chin, R. L.; Hercules, D. M. J. Catal. 1982, 74, 121.

32   Arnoldy, P.; Moulijn, J. A. J. Catal. 1985, 93, 38.

33   Miranda, R.; Chandesris, D.; Lecante, J. Surface Sci. 1983,
     130, 269.

34   Little, J. A.; Butler, G.; Daish, S. R.; Tumbridge, J. A.
     Proc. 8th Intern. Congress Catal.; Verlag Chemie Weinheim,
     1984; Vol. 5, p. 239.

RECEIVED October 18, 1985

In Situ Electron Microscopic Studies of Ni-TiO₂ Interactions

J. A. Dumesic[1], S. A. Stevenson[1], J. J. Chludzinski[2], R. D. Sherwood[2], and R. T. K. Baker[2]

[1] Department of Chemical Engineering, University of Wisconsin, Madison, WI 53706
[2] Corporate Research Science Laboratories, Exxon Research and Engineering Company, Annandale, NJ 08801

A number of investigations have indicated that so-called "strong metal-support interactions (SMSI)" are caused by the migration of partially reduced oxide species onto the surface of titania supported metal particles (1-8). While this is an attractive theory in that it can account for the observed modifications in chemical properties of the metal particles, there is no agreed mechanism by which the postulated transport processes occur.

In the present study we have used two approaches to investigate the respective roles of the oxide support and the metal in this interaction. In the first series of experiments we have used the formation of filamentous carbon as a catalytic probe reaction of the ability of reduced titania species to spread on the surface of nickel. It is well known that when bulk nickel or supported nickel particles are exposed to a hydrocarbon environment at temperatures in excess of about 920 K, the metal particles catalyze the growth of filamentous carbon (9). Figure 1a is an electron micrograph showing the formation of carbon filaments on a nickel grid which had been cooled to room temperature after reaction in acetylene at 1020 K. A schematic representation of the carbon flow path through an individual metal particle is given in Figure 1b. During the filament growth sequence, the catalyst particle is detached from the bulk metal surface as a result of precipitation of dissolved carbon at the rear of the particle.

Inhibition of filamentous carbon growth on metal surfaces by addition of titanium oxide has been reported previously (10). In these experiments the titanium oxide was in the fully oxidized state and was an effective physical barrier toward filament formation, provided that the temperature was kept below 920 K. Above this temperature, the oxide tended to spall, and as bare metal was exposed to the gas phase, prolific filament growth occurred. In the present study we have combined electron microscopy studies with

0097-6156/86/0298-0099$06.00/0

a

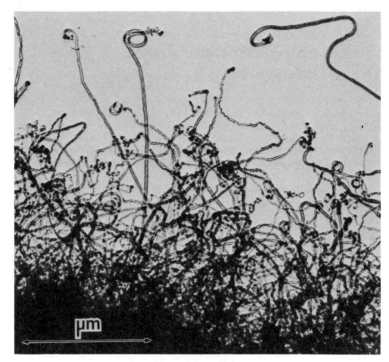

b

Figure 1   (a) Typical appearance of carbon filaments produced on the surface of nickel after exposure to 2
Torr acetylene at 1020 K. (b) Schematic representation of the carbon filament growth on a
nickel particle.

macro-scale experiments performed in a flow reactor, where the
weight of carbon deposited on the specimen was monitored as a function of temperature. Both types of experimental procedures indicated that following reduction in hydrogen at 770 K, titanium oxide
is an extremely effective inhibitor for carbon deposition from
hydrocarbons up to temperatures of at least 1120 K, whereas following oxidation in oxygen, the additive has an almost negligible
effect at 1120 K. This inhibition effect after reduction is
rationalized in terms of the ability of reduced titanium oxide to
wet and spread on the surface of the nickel, preventing the growth
of filamentous carbon which would normally take place when the hot
metal surface was exposed to a hydrocarbon environment. As such,
we have discovered a novel method of inhibiting catalytic carbon
deposition on nickel surfaces.

In a second set of experiments, overlapping thin films of
nickel and titania were heated to 1100 K in the presence of hydrogen in a controlled atmosphere scanning transmission electron
microscope (STEM). During reaction the metal film was observed to
restructure into discrete particles which eventually moved and
attacked the titania substrate. We believe that this behavior
provides direct evidence that both the mobility of reduced titania
and metal species are involved in the initiation of strong metal-
support interactions.

EXPERIMENTAL

(a) Carbon Deposition Studies on Nickel-Titanium Oxide
The changes in appearance of nickel specimens undergoing reaction with hydrogen, oxygen and acetylene were monitored continuously by controlled atmosphere electron microscopy. Sections of
nickel foil were spot-welded across a hole in a platinum heater
strip and sprayed with a solution containing dispersed titanium
oxide (Degussa P-25) in butanol. With this arrangement it was
possible to observe reactions occurring at the edges of the nickel
by following the changes in shape of the sample silhouette as a
function of time and temperature.

The microscopy experiments were complemented by macro-scale
studies performed in a flow reactor. Here the weight of carbon
deposited on treated nickel foils was measured after reaction for 1
h in 1 atm ethane at 870, 970 and 1070 K.

(b) Interaction of Thin Films of Nickel and Titanium Oxide
In situ electron microscopy studies were performed with overlapping thin films of nickel and titanium dioxide. A nickel film
approximately 30 nm thick was prepared by vacuum deposition of the
metal onto a single crystal of sodium chloride. The salt was dissolved in distilled water, and the residual metal film was

thoroughly washed and picked up on a nickel grid. Titania films of similar thickness were grown by oxidizing the surface of a titanium metal foil in oxygen at 620 K; the metallic titanium substrate was subsequently removed in an aqueous solution of 10% nitric and 2% hydrofluoric acids (11). Sections of the oxide film were then floated onto the previously mounted nickel films to create the specimen geometry depicted in Figure 2, i.e., (A) a region of Ni, (C) a region of TiO_2 and (B) a zone of direct contact between Ni and TiO_2.

These specimens were then examined in a JEOL 200 CX electron microscope equipped with an environmental cell and heating stage, which allowed specimens to be heated in 1 Torr hydrogen up to 1100 K; under these conditions the spatial resolution was 0.8 nm. In situ chemical analysis of the reacting specimen could be obtained using electron diffraction, energy-dispersive X-ray analysis, and electron energy loss spectroscopy.

Post-reaction STEM examinations were also performed on a series of nickel-titanium oxide specimens which had been treated for 1 h in a flow reactor in 10% hydrogen/argon at atmospheric pressure and temperatures from 970 to 1120 K. Prior to removal from the reaction zone, specimens were passivated by cooling to room temperature in flowing argon followed by exposure to a flow of 2% carbon dioxide-argon for 1 h. As will be demonstrated later, careful passivation is necessary to ascertain the corresponding chemical state of the specimen from a post-reaction examination.

The gases used in this work, hydrogen, oxygen, argon, acetylene and ethane, were obtained from Scientific Gas Products, Inc. with stated purities of >99.5% and were used without further purification.

RESULTS AND DISCUSSION

(a) Carbon Deposition Studies on Nickel-Titanium Oxide

The titanium oxide, which was initially dispersed on nickel via butanol solution, was readily detectable as isolated ragged aggregates of particles against the smooth uniform nickel edges. No changes in appearance of the specimen silhouette were observed in 1 Torr hydrogen until the temperature was raised to 770 K. At this stage the morphology of the oxide aggregates first changed from a ragged to a smooth outline, followed by a decrease in particle height and a corresponding increase in particle width, suggestive of spreading over the nickel surface (transition A to B in Figure 3). In general, the smaller particles appeared to exhibit this transformation first. As the temperature was gradually increased to 870 K, the oxide spread to such an extent that it was impossible to locate the position of some of the original titania aggregates.

When these samples were subsequently treated in 5 Torr acetylene, they did not display the behavioral pattern characteristic of

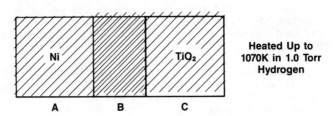

Figure 2   Specimen geometry of overlapping films of nickel and titanium oxide.

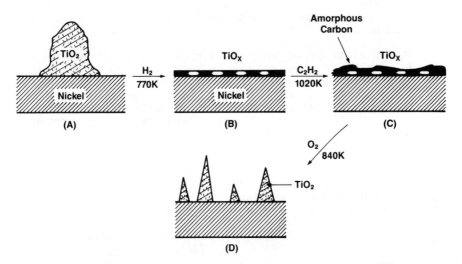

Figure 3   Behavioral pattern of nickel specimens coated with titanium oxide following sequential treatments in hydrogen, acetylene and oxygen.

clean nickel foils (see Figure 1). Although there were some isolated regions of the surface where sparse accumulations of amorphous carbon were detected at 1020 K, over 95% of the surface remained unchanged (Point C in Figure 3). This condition persisted
even when the temperature was raised to 1120 K. An important finding was that if the initial pretreatment in hydrogen was performed
for 1 h at temperatures near 570 K instead of at 770 K, then no
morphological transformation of the titanium oxide occurred and
mitigation of filamentous carbon growth was not achieved to any
significant extent when specimens were subsequently reacted in
acetylene.

In some experiments the ostensibly inert samples (Point C in
Figure 3) were cooled to room temperature and the hydrocarbon replaced by 5 Torr oxygen. Upon reheating, the smooth rippled surface was seen to undergo a dramatic change at 840 K, restructuring
into a saw-toothed form (transition C to D in Figure 3). If these
samples were reheated in acetylene, prolific carbon filament formation ensued at 870 K. If, however, an intermediate reduction step
in hydrogen at 870 K was employed, then the surface structure re-
verted to its smooth rippled form and showed no tendency to catalyze filament formation when subsequently treated with acetylene at
temperatures up to 1120 K.

The amount of carbon deposited on nickel foils following reaction in ethane at atmospheric pressure and temperatures from 870 to
1070 K is presented in Table 1. Examination of these data shows
that titanium oxide provides an effective barrier towards carbon
deposition at 870 K with or without hydrogen pretreatment; however,
at higher temperatures only titanium oxide reduced in hydrogen
exhibits an inhibiting effect.

TABLE I

Amount of carbon deposited on nickel specimens which have been
reacted at various temperatures in 1 atm ethane for 1 h

Specimen	Weight of Carbon Deposited $(g/10^{-4} cm^{-2})$		
	870 K	970 K	1070 K
Virgin Nickel	1.9	26.2	27.7
Nickel/Titanium Oxide	0.2	24.5	25.0
Nickel/Reduced Titanium Oxide	0.2	10.0	6.4

These results provide direct evidence that reduced titania
species migrate over nickel at 770 K under reducing conditions;

this change in morphology of the oxide particles coincides with conditions where strong metal-support interactions are induced for nickel supported on titanium oxide (e.g., 12,13). Furthermore, the restructuring of the reduced titanium oxide from a highly wetting to nonwetting configuration during treatment in oxygen at 840 K correlates with the restoration of normal chemisorption properties of a metal/titanium oxide system, initially in the SMSI state, after exposure to oxygen at 870 K (14).

(b)   Interaction of Thin Films of Nickel and Titanium Oxide
When the specimens consisting of nickel and titania films were heated in 1 Torr hydrogen, regions containing nickel restructured at 800 K to form large particles (10-400 nm in size) supported on an extremely thin nickel film. On continued heating to 1000 K, light symmetrical shaped halos were observed to develop around the nickel particles located on the titanium oxide film (Zone B in Figure 2) suggesting that the oxide was becoming thinner in these regions. This interaction became more extensive as the temperature was increased to 1070 K, as seen from the series of micrographs (Figure 4) taken from the T.V. monitor of the video recorder used to follow this process. This sequence provides direct evidence that nickel particles can move on titanium oxide under these conditions, as seen from the changing position of the indicated particles. Furthermore, during this process, the mobile metal particles remove material from the oxide support, and as a consequence create unsymmetrical pits in the titania film (see Figure 5). These results also indicate that the metal particles only attack those regions of the support with which they are in contact. Since the size of the pits that are produced in the oxide are in general larger than the metal particles, it is possible that a fraction of the titania is transported inside the particles or volatizes during the reaction.

TABLE II

Electron diffraction analysis of titania after reaction in hydrogen at 1070 K in the presence of nickel:

d-spacings from diffraction pattern (nm)	d-spacings from crystal structure of Ti_4O_7 (nm)
0.427	0.427
0.332	0.338
0.308	0.302
0.276	0.282, 0.280
0.258	0.263
0.248	0.252, 0.248

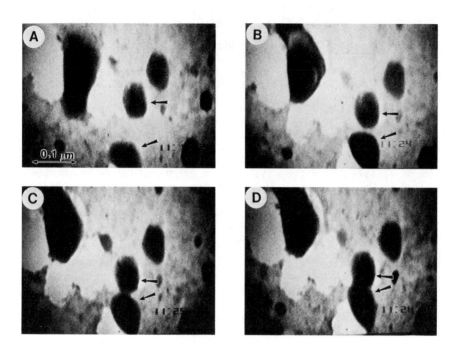

Figure 4   Sequence showing the interaction of nickel particles with titanium oxide in 1 Torr hydrogen at
1070 K.

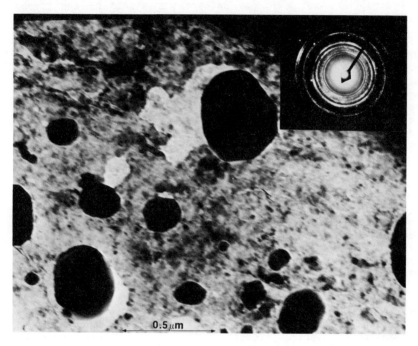

Figure 5   Electron micrograph showing pits in titanium oxide
support produced by nickel particles during reaction in 1 Torr hydrogen at 1070 K.   Inset shows
the electron diffraction pattern of the support
under these conditions.

The extent of reduction of the titania support was found to be a function of the reaction temperature. In situ electron diffraction analysis showed that the oxide underwent transformation from TiO_2 to Ti_4O_7 at 1070 K (see inset of Figure 5 and Table II). This pattern was not found at lower temperatures; however, it is possible that Ti_4O_7 was formed in the vicinity of the nickel under these conditions but that its pattern was masked by the strong rutile signal.

Post-reaction examination of similarly prepared specimens that had been treated in a flow reactor at 1 atm in 10% H_2/Ar at 970 K showed many of the features observed during the in situ studies. A sample treated in the flow reactor at 970 K looked comparable to a sample that had been treated in 1 Torr H_2 at 1070 K in the microscope, i.e., many of the nickel particles had unsymmetrical pits associated with them. Reduction at 1120 K in the flow reactor produced dramatic changes in the appearance of samples, with extensive pitting of the titania film observed.

The electron diffraction patterns of these samples paralleled those observed for the samples reacted in the electron microscope, i.e., samples reduced in the flow reactor at 1120 K showed that the titanium oxide had been completely converted to Ti_4O_7, whereas samples treated at 970 K showed only the rutile form of TiO_2, in agreement with the work of Bell and coworkers ([15]). One anomalous finding occurred with a sample that was treated at 1120 K for which the passivation procedure was inadvertently omitted; the diffraction pattern showed no evidence of Ti_4O_7.

Finally, one may ask why the attack of titania by nickel (e.g., formation of pits) has not been observed in previous studies where nickel/titanium oxide samples had been treated in hydrogen and examined by electron microscopy ([6,12,13]). In these previous cases, the nickel particles were considerably smaller (5 to 10 nm) than the ones formed in the present experiments (~100 nm) and therefore the extent of reaction may have been only sufficient to cause removal of a few monolayers of titania, which would be difficult to detect.

SUMMARY

The results of the present study demonstrate that migration of both titania and metal species may be involved in the initiation of strong metal-support interactions during treatment in hydrogen at temperatures in excess of 770 K.

Literature Cited

1. Engels, S., Freitag, B., Mörke, W., Roschke, W., and Wilde, M., Z. Anorg. Allg. Chem. 474, 209 (1981).

2. Meriaudeau, P., Dutel, J. F., Dufaux, M., and Naccache, C., Stud. Surf. Sci. Catal. 11, 95 (1982).
3. Santos, J., Phillips, J., and Dumesic, J. A., J. Catal. 81, 147 (1983).
4. Jiang, X-Z, Hayden, T. F., and Dumesic, J. A., J. Catal. 83, 168 (1983).
5. Resasco, D. E. and Haller, G. L., J. Catal. 82, 279 (1983).
6. Simoens, A. J., Baker, R. T. K., Lund, C. R. F., Dwyer, D. J. and Madon R. J., J. Catal. 86 359 (1984).
7. Sadeghi, H. R. and Henrich, V. E., J. Catal. 87, 279 (1984).
8. Chung, Y-W., Xiong, G., and Kao, C-C., J. Catal. 85, 237 (1984).
9. Baker, R. T. K., Barber, M. A., Harris, P. S., Feates, F. S., and Waite, R. J., J. Catal. 26, 51 (1972).
10. Baker, R. T. K., and Chludzinski, J. J., Jr., J. Catal. 64, 464 (1980).
11. Tatarchuk, B. J., and Dumesic, J. A., J. Catal. 70, 3 (1981).
12. Mustard, D. G., and Bartholomew, C. H., J. Catal. 67, 186 (1981).
13. Smith, J. S., Thrower, P. A., and Vannice, M. A., J. Catal. 68, 270 (1981).
14. Baker, R. T. K., Prestridge, E. B., and Garten, R. L., J. Catal. 59, 293 (1979).
15. Singh, A. K., Pande, N. K., and Bell, A. T., J. Catal., in press.

RECEIVED September 17, 1985

11

Presence and Possible Role of Pt Ions in Hydrocarbon Reactions

M. J. P. Botman, Li-Qin She, Yia-Yu Zhang, T. L. M. Maesen, T. L. Slaghek, and V. Ponec

Gorlaeus Laboratories, State University of Leiden, P.O. Box 9502, 2300 RA Leiden, The Netherlands

Unreducible ions of the transition metal component
of supported catalysts, which survive a severe reduction, are clear evidence of a strong interaction
between the support and the metallic element.
Currently there is a discussion in the literature
on this subject, specifically: 1) Do some Pt-ions
survive a severe reduction? 2) Are they accessible?
3) Do they play a role in the metal-support interaction and reactions of hydrocarbons?
Results presented here allow a positive answer to
the first two questions. The third question is
complex and cannot be answered definitely.

The problem of the interaction between a catalytically active metal
and its support is not new in the catalytic literature but during
the last decennium the interest in this problem has increased.
Research in this field has been stimulated in a quite remarkable
way by several papers ($\underline{1-4}$) and in a short time a great volume of
information has accumulated ($\underline{5}$).
   Interesting phenomena observed gave rise to some speculations
which have lacked sound physical justification; examples are the
speculations on a massive transfer of electrons from (or to) the
support to (or from) the metal, the ideas about changes of the work
function of the metal by such a transfer or by the field of the support, etc. However, these ideas have received considerable attention
and unfortunately they have tended to shadow other pieces of information concerning the metal-support interaction.
   One of the reliable indications of a real strong interaction
between a support and the metallic component of a catalyst is the
presence, after a severe reduction, of unreduced ions of the metallic component. For example, there is an extensive literature in
this respect on Ni and Cu and some other catalysts ($\underline{6-11}$) demonstra-
ting the presence of Ni and Cu ions in reduced catalysts. However,
with Pt the situation is less clear.

0097-6156/86/0298-0110$06.00/0

McHenry et al (12) reported at the 2nd Intl.Congress on Catalysis that Pt ions survived a severe reduction and were, according
to the authors responsible for dehydrocyclization and aromatization.
Some workers confirmed this claim (13-15), while some others rejec-
ted this conclusion (16-18). Because of the extreme importance of
such conclusions for catalyst preparation procedures ("can and
should the Pt ions be stabilized against the reduction?") we have
addressed this question and we report below the results of our study.
The questions to be answered are as follows.
i) Does a fraction of Pt^{n+} in the alumina or silicas supported catalysts survive the reduction and is the fraction sufficient to have
potential importance?
ii) Are the Pt^{n+} ions accessible to the gas phase?
iii) Does the presence of the second component in the "bimetallics"
influence the fraction of Pt present as ions?
iv) Do the Pt^{n+} ions manifest themselves in the hydrocarbon skeletal
reactions?
      Owing the difficulties of detecting small amounts of ions in
close proximity to the metal by XPS, an UV/VIS or other spectroscopic techniques (with XPS the interpretation of shifted peaks is not
free of problems, decreases in the particle size shifts the Binding
Energy by the same amount as one positive charge) we decided to use
the classical chemical determination by a selective extraction of
ions.

Experimental

Catalytic tests were performed with n-hexane/H_2 mixtures (1/16) at
atmospheric pressure. The total flow of gases in the feed was
varied between 10 and 20 ml/min. Under these conditions the total
conversion was always < 10%. Analysis of products was performed by
GLC. The column used contained 15% squalane on chromosorb P-AW
with DMCS, the length of the column was 5m. Description of the flow
apparatus and of data evaluation can be found in earlier papers (19).
      One detail of the experimental procedure should be considered
in more detail. Regardless of the procedure used, the catalysts always exhibit some slow decay in the activity and sometimes an extended selfpoisoning can cause changes in selectivities. We observed
that the decay is small and has no influence on the selectivities
when the following procedure is adopted. It is measured from low
to high temperatures and after each temperature jump, the temperature is kept constant for 30 minutes.
      Extraction was performed in a special cell which allowed ad-
mission of oxygen free extracting solution to the reduced samples.
The extraction solution consisted of ethylene diamine and 4%
CH_3COOH. Samples covered by the liquid was placed under the reflux
cooler and the mixture (with the powder) brought to boiling.
Refluxing was continued for about 15 hours and then the powder was
separated by centrifugation and the content of Pt in the solution
determined by Atomic Absorption Spectroscopy. Calibration experiments showed "ionic" Pt could be recovered from supported chlorides
and oxides completely and vice versa and that some reduced samples
containing small Pt particles on inert carrier (SiO_2) has less than
0.5% Pt as ions.

In contrast in systems where ions could be reasonably expected, the
concentration of Pt ions varied between 0.5 and 9% of the total Pt
content. It is important to notice that the presence of CH_3COOH is
essential for complete extraction of Pt^{n+} from oxygenated ligands.
We have also examined some other methods. Extraction by $SnCl_2$ appeared to be entirely unreliable since it cannot extract Pt^{n+} out
of bulk oxide. Extraction by acetylacetone caused difficulties when
the system was rich in chlorine. Therefore, the method as described
above was finally adopted.

Catalysts used were prepared by impregnation and after reduc-
tion they contained different amounts of extractable Pt. The follo-
wing catalysts were tested:
Catalyst A: from Pt dissolved in aqua regia (H_2PtCl_6 in excess
HCl/HNO_3).
Catalyst B: from H_2PtCl_6 dissolved in H_2O.
Catalyst C: from $Pt(NH_3)_4(OH)_2$
Catalyst D: prepared from $H_2Pt(OH)_6$ dissolved partially in hot water
and precipitated finally on the surface of alumina.
All these catalysts were 1% wt Pt/Al_2O_3; Al_2O_3 was Ketjen Alumina
CK300. Catalysts were prepared both with and without a calcination
step (2 hours at 723K, in air).
Standard reduction in situ (of samples as such, or prereduced) was
4 hours at 723K.

Results

Table I shows the region of concentration in which the Pt^{n+} ions
are found to exist after a standard reduction (723K, 4 hours)

Table I.   Content of Pt ions in the Pt/Al_2O_3 reduced catalysts

catalyst	precursor	% Pt ions calcined	not calcined
A	$Pt(NH_3)_4(OH)_2$	0.7	0.8
B	H_2PtCl_6 in H_2O	2.0	1.7
C	H_2PtCl_6 in aqua regia	4.0	3.1
D	$H_2Pt(OH)_6$	0.4	0.6

When a second component was added to the catalyst, the concentration
of Pt^{n+} ions was suppressed, as can be seen from figure 1.
It is important to know whether the Pt ions, if present, are
also accessible to the gas phase. The extracted Pt can, in principle,
also comprise those Pt ions which are present as "anchors" hidden
under the metal particles. One way to establish the accessibility
is to use a test reaction, which is known to occur (solely or pre-
ferrentially) on metal ions. It has been found that the formation
of methanol from CO/H_2 mixture is such reaction ([11,20]). A test by
this reaction showed a positive correlation between the amount of

Pt ions and the activity of this reaction. We shall report on this aspect in more detail elsewhere. Another point of interest concerns the nature of the environment in which unreducible Pt ions embedded. It has been shown from the UV/VIS reflection spectra that Cl ligands of Pt complexes exchange rather easily with the OH (or O?) ligands on the support surface. Nevertheless, after calcination, the UV/VIS spectrum of catalysts made from chlorine containing precursors and of those made from some chlorine free precursors ($H_2Pt(OH)_6$) are different in the region of the spectrum where the ligand to metal charge transfer band usually appears. However, some other chlorine free compounds ($Na_2Pt(OH)_6$) are similar in this respect to a calcined H_2PtCl_6 catalyst. This probably indicates that the Pt ions surviving the reduction are in a different local environment (mainly an oxygen containing environment) in the two mentioned groups of products of calcination.

The test of catalytic behaviour in the n-hexane reactions has been performed in two temperature regions separately.
a) in the "low T region" (up to about 593K), where the metallic function of the supported Pt operates predominantly, and
b) in the "high T region" (653K, up to 773K), where the support also shows an activity.
The low temperature region offers the following picture. At the lowest temperature the selectivity for hydrogenolysis S_H is the highest selectivity with all catalysts studied, and this selectivity decreases with the increasing temperature following a "Z-form" curve. The selectivity for isomerization S_I starts low but increases with increasing temperature, forming an S-shaped curve. The total dehydrocyclization selectivity rises to 100% and it shows usually a featureless (almost a constant) curve. The S_H and S_I cross each other in the region between 533 and 553K. Calcined catalysts, when compared with the uncalcined catalysts of the same composition, showed a cross point at a little higher temperature. However, this behaviour is not correlated with the concentration of Pt ions. When excess Cl is present in the system calcination leads to a higher concentration of ionic Pt, but with other systems calcination did not always increase Pt^{n+} concentration measurably.

As our discussion below will demonstrate, the high temperature region is a more difficult feature to study than the low temperature region. Typical results and problems are demonstarted by figure 2, showing the selectivity in dehydrocyclization S_D as a function of temperature. Selectivity in isomerization S_I varies complementary to the variations in S_D; under conditions where S_D is high – D_I is low, and vice versa. Selectivity in hydrogenolsyis rises to 100% and does not vary significantly in the same region. Data shown in figure 2 were obtained at low and comparable conversions but at different contact times (contact time being characterized by the ratio S/F, with S = metal surface area and F = feed).

Discussion

Two of the three questions formulated in the introduction can now be answered;
1) A measurable amount of Pt ions survives the reduction at 723K; and

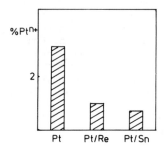

Figure 1. Content of Pt ions in reduced catalysts.
Catalysts: 1% wt Pt/Al$_2$O$_3$, to which an equimolar
amount of Sn or Re has been added.

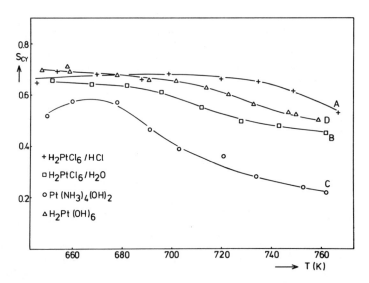

Figure 2. Selectivity in the total dehydrocyclization S$_D$, as
a function of temperature. Standard reaction conditions.
For the codes, see table 1.

2) Some of these ions are certainly accessible to the gas phase.
The third question concerning the role of Pt ions can only be answered for the low temperature region: there are no indications
found for any role of Pt ions in this region. The data show that the
calcination step can, indeed, influence the shape of the selectivities vs temperature curves, however, this effect is not particularly
pronounced.
    With regard to the high temperature region, the study is still
in progress. What follows therefore is merely a progress report on
the state of affairs.
Let us start by reviewing the effects and phenomena which are of
potential importance for the high temperature region. For convenience
we divide them into two classes: I.Intrinsic effects. II. Secondary
and indirect effects. Under I. the following is understood.
a) Pt ions could be, as suggested by McHenry et al (12), the centres
of dehydrocyclization. This possibility should be considered for two
reasons. First, an important mechanism leading to aromatization and
dehydrocyclization (21,22) proceeds via olefines (polyolefines) and
stepwise dehydrogenation. Such reaction does not require a large en-
semble of sites and the (poly)olefinic complexes can be more easily
stabilized on positively charged Pt. Second, the olefines are indeed
detected under the reaction conditions (23).
b) Particle size effects. In principle, some geometric configurations
can be intrinsically better in dehydrocyclization than some others
(24,25).
c) Acidity of the support is varying with different preparations.
    The following is understood under class II of the secondary
and indirect effects.
a) Consecutive reactions.
In the high temperature region consecutive reactions play a role also
when the overall conversion is kept low (under 10%); a fraction of
isomerization products are formed through a ring opening of the previously produced methylcyclopentane (when hexane is the feed). This
indicate that each shortening of the apparent contact time τ enhances
dehydrocyclization. Contact time τ can be varied in a controlled
manner by changing the amount of the catalyst and the extent feed.
The upper limit of variations being controlled by the extent of
conversion, that has to be kept low. However, τ is also sensitive
to effects which cannot easily be controlled. For example, when the
selfpoisoning is stronger with a certain catalyst, that catalyst
functions under an effectively shorter τ and can as a result exhibit
a higher dehydrocyclization selectivity.
b) It is known that Pt particles of larger size are more easily self-
poisoned by the carbon(aceous) layer   formation, then the smaller
particles, which are less susceptible to selfpoisoning (26). This
feature leads to an enhanced dehydrocyclization also with Pt on inert silica (26) and, of course the enhancement can be intrinsic or
induced by the carbon(aceous) layer (in one way or another). For
example, by changing the effective τ. Particle size can thus have
direct effects (as mentioned under I, b) or indirect effects as just
described.
    From figure 2, it is apparent that the differences between the
catalysts are most pronounced in the region of the highest temperatures, conditions where the consecutive reactions are expected to

become more important. When the catalysts A, B and D are compared at approximately the same contact time τ (i.e. approximately the same metal surface area of the catalyst in the reactor) then with comparable overall conversions the sequence in selectivity towards cyclization of these catalysts is changed. This strongly suggests that some of the differences in figure 2 are due to consecutive reactions.

Two catalysts with the highest and lowest concentration of ions (but of different particle sizes) show the highest dehydrocyclization activity. This leads to the conclusion that the role of ions (if any) is rather limited or masked by compensation due to differences on other effects mentioned above.

Conclusions

1) A small amount of Pt ions survives on Al_2O_3 after a severe reduction treatment and can be suppressed by the presence of another metal.
2) A part of these ions is accessible.
3) At low temperature these ions have little influence on the skeletal reactions of hydrocarbons. A possible role of ions in the high temperature region is not excluded but as yet is not proven.

Literature cited

1. Dalla Bata, R.A.; Boudart, M. In "Catalysis"; Hightower, J.W.,
    Ed; Proc.5th Int.Congr.on Catal., Miami Beach, 1972, North
    Holland, Elsevier, 1973, Vol.2, p.1329
2. Bond, G.C.; Sermion, P.A. Gold Bulletin 1973, 105
3. Tauster, S.J.; Fung, S.C. and Garten, R.L. J.Am.Chem.Soc.
    1978, 100, 170
    Tauster, S.J. and Fung, S.C. J.Catal. 1978, 55, 29
4. Tauster, S.J.; Fung, S.C.; Baker, R.T.K. and Horsley, J.A.
    Science 1981, 211
5. Studies in Surface Science and Catalysis II. "Metal–Support
    and Metal.Additive Effects in Catalysis", Imelik et al, eds.;
    Elsevier 1982
6. Montes, M.; Penneman de Bosscheyde, Ch.; Hodnett, B.K.;
    Delannay, F.; Grange, P. and Delmon, B. Applied Catal. 1984,
    12, 309
7. Schuit, G.C.A.; de Boer, N.H. Rec.Trav.Chim., Pays-Bas 1951,
    70, 1067; 1953, 72, 909
8. de Jong, K.P.; Geus, J.W. and Joziasse, J. Applied Surf.Sci.
    1980, 6, 273
9. Schats, W.M.T.M. Dissertation Thesis, Catholic University,
    Nijmegen, The Netherlands, 1981
    Coenen, J.W.E.; In "Preparation of Catalysts II", Demon, B.;
    Grange, P.; Jacobs, P. and Poncelet, G., eds., Elsevier,
    1979, p.89
10. Lojacono, M. and Schiavello, M. In "Preparation of Catalysts",
    Delmon, B.; Jacobs, P.A. and Poncelet, G., eds., Elsevier,
    1976, p.473

11. Driessen, J.M.; Poels, E.K.; Hindermann, J.P. and Ponec, V. J.Catal. 1983, 82, 26
12. McHenry, K.W.; Bertolacini, R.J.; Brennan, H.M.; Wilson, J.L. and Seelig, H.S., Actes du 2ieme Congr.Intl.de Catalyse, Paris 1960, Editions Techniques, Paris 1961, Vol.2 p.2295
13. Bursian, N.R.; Kogan, S.B.; Davydova, Z.A. Kinet.and Catal. 1967, 8, 1085
14. Johnson, M.F.L. and Keith, C.D. J.Phys.CHem. 1963, 67, 200
15. Putanov, P.; Jovanovic, M. and Selakonic, O. React.Kinet. Catal.Lett. 1978, 8(2), 223
16. Shekhobalova, V.I. and Lukyanova, Z.V. Zhur.Fiz.Kchim (Russ.J.Phys.Chem.) 1979, 53, 1551
17. Lietz, G.;Lieske, H.; Spindler, H.; Hanke, W. and Völter, J. J.Catal. 1983, 81, 17
18. Kluksdahl, H.E. and Houston, R.J. J.Phys.Chem. 1961, 65, 1469
19. Ponec, V. and Sachtler, W.H.M. In "Catalysis"; Hightower, J.W. Ed.; Proc.5th Intl.Congr.on Catalysis, Miami Beach, 1972, North-Holland, Elsevier, Amsterdam, 1973, Vol.2, p.645 van Senden, J.G.; van Broekhoven, E.H.; Wreesman, C.T.J. and Ponec, V. J.Catal. 1984, 87, 468
20. van der Lee, G.; Schuller, B.; Post, H.; Favre, T.L.F. and Ponec, V. J.Catal. submitted
21. Finlayson, O.E.; Clarke, J.K.A. and Rooney, J.J. J.Chem.Soc. Faraday Trans.I 1984, 80, 191
22. Ponec, V. Adv.Catalysis 1983, Vol.32, 149
23. van Broekhoven, E.H. and Ponec, V. J.Molec.Catal. in print
24. Davis, S.M.; Zaera, F. and Somorjai, G.A. J.Catal. 1982, 77, 439; 1984, 85, 206
25. Somorjai, G.A. Proc.8th. Intl.Congr.on Catalysis, Berlin, 1984 Dechema BRD, 1984, Vol.1, p.113
26. Lankhorst, P.P.; de Jongste, H.C. and Ponec, V. In "Catalyst Deactivation"; Delmon, B. and Froment, G.F. eds.; Elsevier, Amsterdam, 1980, p.43

RECEIVED November 13, 1985

12

Support Effect on Chemisorption and Catalytic Properties of Noble Catalysts

P. Mériaudeau, M. Dufaux, and C. Naccache

Laboratoire Propre du C.N.R.S., Institut de Recherches sur la Catalyse, Conventionné à l'Université Claude Bernard, Lyon I, 2, Avenue Albert Einstein, 69626 Villeurbanne Cédex, France

Although the most well established function of the support is to disperse the metal and to retard the sintering, recent works on metal-support interaction collected in (1) have shown that the chemisorptive and catalytic activity of group VIII metals were considerably lowered when the metal was supported on reducible oxide, such as TiO_2, and reduced at high temperature. The objective of the work reported in this paper was to further investigate the role of the support on the adsorptive and on the catalytic properties of platinum.

Results and discussion

Supported-Pt-catalysts were prepared by wet impregnation technique using H_2PtCl_6 solution. Supported platinum salt was reduced at 473 K and 773 K, H_2, CO adsorption and electron microscopy were used to measure the metal dispersion. On table 1 are reported Pt dispersion, infrared ν_{CO} for CO adsorbed on Pt and the catalytic data on benzene hydrogenation and n-hexane conversion.

As already reported (2) the adsorption of H_2 and CO is greatly lowered for Pt-TiO_2 reduced at 773 K. The decrease of hydrogen adsorption capacity is accompanied by a decrease of the catalytic activity for benzene hydrogenation and n-hexane hydrogenolysis. However there is no change in ν_{CO} stretching frequency of adsorbed CO. In contrast for Pt-CeO_2 reduced at 773 K only a small decrease of H_2-adsorbed was observed. Furthermore the ν_{CO} frequency was lowered by 16 cm^{-1}.

0097-6156/86/0298-0118$06.00/0

TABLE 1

Catalyst	T_{red}	H/Pt_s	ν_{CO} cm^{-1}	$r_{C_6H_6}$ a	$r_{C_6H_{14}}$ b
Pt-TiO$_2$	473 K	0.25	2080	1	75
4.8 wt % Pt	773 K	0.05	2080	0.08	2
Pt-CeO$_2$	473 K	0.6	2088	1	1119
1.8 wt % Pt	773 K	0.4	2072	0.2	173

a : relative rate r_T/r_{473} ; b : rate n-hexane in m.moles $h^{-1}g_{Pt}^{-1}$

Electron microscopy studies indicated that the Pt particle sizes remained almost unchanged between low and high temperature reduced samples. n-hexane conversion on Pt-TiO$_2$ and Pt-CeO$_2$ reduced at 473 and 773 K was for both catalysts considerably lowered for those samples reduced at high temperature. However on Pt-TiO$_2$ the strong decrease in the reaction rate occured without any significant change in the product distribution (about 20 % cracked product C_1-C_5) while on Pt-CeO$_2$ reduced at 773 K n-hexane hydrogenolysis is considerably reduced compared with Pt-CeO$_2$ reduced at 473 K (48 % cracked products on Pt-CeO$_2$ 473 K and 17 % cracked products on Pt-CeO$_2$ 773 K). Both TiO$_2$ and CeO$_2$ form n-type semiconductor when reduced at high temperature. It results that both TiO$_2$ and CeO$_2$ reduced at high temperature would have the ability to donate electrons to the supported platinum particles. This model would lead to the same effect for both supports in contrast with our results (3). We conclude that in the case of TiO$_2$ the reduction of H$_2$, CO adsorption and consequently the decrease in the catalytic activity of Pt is mainly due to a coverage of Pt surface by a suboxide TiO$_x$ layer (3,4) which upon oxygen adsorption is destroyed. The mobility of reduced cerium oxide could be less. However cerium cations were reduced to cerium zerovalent which at high temperature would form Pt-Ce intermetallic compounds which explains the shift in ν_{CO} as well the change in benzene hydrogenation and n-hexane conversion.

Support effect on the synthesis of methanol over Pt

It has been reported that methanol is formed from CO-H_2 reaction over silica-supported Pd, Pt, Ir catalysts (5). The behaviour of the metal was found to be influenced by the carrier (6,7,8,9). The selectivity in methanol was discussed in terms of acid-base properties of the support which influenced the non dissociative adsorption of CO on the metal required for oxygenated hydrocarbon formation, in terms of electronic interaction between the metal and the support or in terms of stabilization by the carrier of oxidized metal cations which would adsorb CO non dissociatively. We have studied the CO-H_2 reaction at 553 K, 30 atmospheres over Pt supported on a variety of oxides. The characteristics of the catalysts are given in table 2

TABLE 2

Catalyst	wt % Pt	H/Pt$_s$	T.E.M. nm	A ^{a}	S_{MeOH} %	S_{ROH} ^{b}
Pt-SiO$_2$	1.8	0.65	2.0	1	46	53
Pt-Al$_2$O$_3$	2.5	0.77	1.5	4.40	26	48
Pt-TiO$_2$	4	0.26	-	6.6	64	67
Pt-ThO$_2$	3.3	0.25	-	8.5	72	77
Pt-CeO$_2$	2.4	-	-	17.7	82	85
Pt-MgO	2.5	0.16	2-6	21.7	95	97
Pt-La$_2$O$_3$	2	-	2.0	26	95	97

a) Activity (MeOH + EtOH) 10^{-2} mole h^{-1}g$_{Pt}^{-1}$; b) Selectivity in oxygenated hydrocarbons (MeOH, EtOH, DME).

The catalytic activity of the supports in CO + H_2 was also investigated. The results indicated that CeO$_2$, ThO$_2$, La$_2$O$_3$ have appreciable activity for methanol formation. Al$_2$O$_3$, MgO and TiO$_2$ have very low activity for MeOH formation.

Table 2 shows clearly that the activities of supported Pt catalysts and to a less extend the selectivities for oxygenated compounds form $CO-H_2$ are strongly affected by the support, the more basic supports having the better promoting effect for methanol formation. Furthermore it is clear that the highest rates of MeOH formation were obtained for Pt supported on the carriers exhibiting a non negligible activity. Mechanical mixture of Pt-support plus support were tested in the $CO-H_2$ reaction. The results are listed in table 3.

TABLE 3

Catalyst	Promoting effect [a]	Selectivity % MeOH
$Pt-CeO_2 + CeO_2$	2.30	80
$Pt-ThO_2 + ThO_2$	3.53	76
$Pt-TiO_2 + CeO_2$	3.6	60
$Pt-SiO_2 + CeO_2$	2.4	74

a) derived from the ratio : experimental MeOH/calculated MeOH which would be given by each component.

The activity of supported Pt catalysts for methanol synthesis from $CO-H_2$ is considerably enhanced when the metal is supported on oxides which exhibit themselves appreciable activity for MeOH synthesis. Furthermore it is found that the rate of methanol formation on Pt-supported catalyst is increased when ThO_2, CeO_2 were mechanically mixed with the Pt catalyst. Such behaviour is typical for bifunctional catalysts. It has already been reported that ThO_2, CeO_2 adsorb carbon monoxide without dissociation. Such activated CO can be hydrogenated to form a formyl species, the formyl species interacting with lattice oxygen will produce a formate intermediate. CO could also react with OH group to form a formate. Formyl or formate intermediates have often been suggested in the mechanism which describes the formation of CH_3OH from $CO-H_2$. If such mechanism prevails for Pt-supported catalyst one could suggest that on $Pt-CeO_2$, $Pt-ThO_2$, $Pt-La_2O_3$, Pt-MgO, the catalyst behaves as a dual site material : CO activated on the oxide is hydrogenated by hydrogen activated on platinum. Such mechanism does not exclude the possibility that CH_3OH could be formed also through CO and H_2 both activated on the metal. However since Pt is also active for CO dissociation with the subsequent formation of methane, $CO-H_2$ reaction occuring only on Pt would give a lower selectivity in CH_3OH. Dual site mechanism on mechanical mixture of supported platinum with oxides would also prevail provided that the pure oxide is active for CO activation and that activated hydrogen could spillover at the solid interphase.

Literature Cited

1. Metal-support and metal-additive effects in Catalysis (eds B. Imelik et al) Elsevier, Amsterdam 1982.

2. Tauster, S.J., Fung, S.C. and Garten, R.L., J. Amer. Chem. Soc., 100, 170, 1978.

3. P. Mériaudeau, J.F. Dutel, M. Dufaux, C. Naccache "Metal-support and metal-additive effects in Catalysis" (eds B. Imelik et al) Elsevier, Amsterdam 1982, p. 95.

4. a) J. Santos, J. Phillips and J. Dumesic, J. Catal., 81, 147 (1983).
   b) D.E. Resaco and G. Haller, J. Catal., 82, 279, 1983.
   c) H.R. Sadeghi and V.E. Henrich, J. Catal., 87, 279, 1984.
   d) A.J. Simoens, R.T.K. Baker, D.J. Dwyer, C.R.F. Lund and R. Madon, J. Catal., 86, 359, 1984.
   e) D.N. Belton, Y.M. Sun, and J.M. White, J. Phys. Chem. 88, 5172, 1984.

5. Poustma, M.L., Elek, L.F., Ibarbia, P.A., Risch, A.P., and Rabo, J.A., J. Catal., 52, 157, 1978.

6. Ryndin, Yu. A., Hicks, R.F., and Bell, A.T., J. Catal., 70, 287 (1981).

7. Poels, E.K., Koolstra, R., Gens, J.W., and Ponec, V., "Metal-support and Metal-additive Effects in Catalysis" (eds B. Imelik et al) Elsevier, Amsterdam 1982, p. 233.

8. Fajula, F. Anthony, R.G., Lunsford, J.H., J. Catal., 73, 237, 1982.

9. P. Mériaudeau, M. Dufaux and C. Naccache, VIII th Congress on Catalysis, Berlin 1984, Vol. II, p. 185.

RECEIVED September 12, 1985

Metal–Support Interactions in Ni Catalysts

A Comparative Study of Kinetic and Magnetic Behavior Between Nb$_2$O$_5$ and Phosphate Supports

E. I. Ko[1], J. E. Lester[2], and G. Marcelin[2,3]

[1] Department of Chemical Engineering, Carnegie-Mellon University, Pittsburgh, PA 15213
[2] Gulf Research & Development Company, Pittsburgh, PA 15230

Different mechanisms of metal–support interactions were found from kinetic and magnetic studies over nickel catalysts on niobia– and phosphate-containing supports. All the catalysts showed a suppression in hydrogen chemisorption after they had been reduced at high temperatures. However, the niobia–containing samples as a group behaved very differently from the phosphate–containing samples in ethane hydrogenolysis and carbon monoxide hydrogenation. When reduced at still higher temperatures, the phosphate–containing catalysts exhibited a change from the normal magnetization–temperature behavior, but the niobia-containing catalysts did not. Instead, the latter showed an apparent loss of nickel metal and a decrease in crystallite size. Furthermore, the extent of interaction in the niobia–containing catalysts could be moderated by using a niobia–silica surface phase oxide.

The work of Tauster and coworkers ([1,2]) showed that hydrogen chemisorption is suppressed on group VIII metals supported on a series of oxides after these samples have been reduced at high temperatures. The term strong metal–support interactions (SMSI) was introduced to describe this behavior. A similar suppression in hydrogen chemisorption has since been reported for many other supported metal systems ([3–5]). However, the use of other chemical probes ([4,5]) demonstrated that different mechanisms of metal–support interactions could exist for different types of oxides. Furthermore, even for a so-called SMSI oxide, the degree of interaction could be influenced by many parameters such as crystallite size and reduction temperature. It would thus be desirable to find an approach to systematically compare catalytic behavior of different systems.

[3] Current address: Department of Chemical and Petroleum Engineering, University of Pittsburgh, Pittsburgh, PA 15261

0097–6156/86/0298–0123$06.00/0
© 1986 American Chemical Society

As an attempt in this direction, a hierarchy was recently developed for nickel catalysts (6). The basic idea is to monitor the
chemical properties of a catalyst as probed by hydrogen chemisorption, ethane hydrogenolysis, and carbon monoxide hydrogenation.
The hierarchy, originally developed for Ni/Nb_2O_5 catalysts, was
later extended to nickel supported on phosphate-containing materials and a niobia–silica surface phase oxide. In this paper the
usefulness of the hierarchy will be illustrated by its ability to
differentiate between support effects of niobia and phosphate, and
to establish the intermediate degree of interaction of niboia-
silica.

Magnetic measurements, which included magnetization–temperature behavior and particle size determination, were also made on
this series of catalysts as a function of reduction treatment.
These results, in conjunction with those obtained from kinetic
studies, provided a physical picture of the different mechanisms
for the niobia and phosphate supports. The same picture is consistent with the less interacting nature of niobia–silica, which
should prove useful as a model system for the study of metal-
support interactions in general.

Experimental

Supports and Catalysts.   The preparation of the supports used in
this study was discussed in detail elsewhere. The two phosphate
supports, $Al_2O_3 \cdot 2AlPO_4$ and $4MgO \cdot 13Al_2O_3 \cdot 10AlPO_4$ were co–precipitated using the necessary nitrate salts, phosphoric acid, and
ammonium hydroxide at a fixed pH (7). Niobia was precipitated by
adding ammonium hydroxide to a methanolic solution of niobium
chloride (8). The niobia–silica support was prepared by impregnating SiO_2 (Davison 952) to incipient wetness with a hexane solution
of niobium ethoxide. The sample was then dried and calcined to
obtain a homogeneous surface phase oxide (9).

Table I summarizes the characteristics of nickel catalysts
prepared onto these supports. For brevity these catalysts will be
referred to by a notation in the form αA-β.   For   example,
7AAP-573 represents a 7 wt % Ni catalyst supported on Al_2O_3
$\cdot 2AlPO_4$ reduced at 573 K for 1 h. Incidentally, this sample did
not reduce under these conditions and was excluded from further
kinetic studies. Notations for the other catalysts are shown in
the first column of Table I. All samples were reduced at the
specified temperature for 1 h unless noted otherwise. The percent
reduction was determined by measuring oxygen uptake at 673 K in a
commercial thermogravimetric system (Cahn 113). The average particle size was determined by either X-ray diffraction line broadening or magnetic measurements (see below).

Chemical Measurements.   All the catalysts in Table I showed a
suppression in hydrogen chemisorption to varying extent. Such a
suppression manifested itself either as an overestimation of crystallite size from chemisorption data or as a lower adsorption
stoichiometry of $H/Ni_{(s)}$ than what would be expected of a comparable Ni/SiO_2 catalyst. In addition, the suppression was more
severe with increasing reduction temperature (8-10). As mentioned

Table I. Summary of Supported Nickel Catalysts

Notation	Support	Preparation	Loading (wt. %)	Reduction	% Reduction	Average particle diameter (nm)
7AAP-573	$Al_2O_3 \cdot 2AlPO_4$	ion exchange with nickel nitrate	7	573K, 1 h	0	—
7AAP-773	$Al_2O_3 \cdot 2AlPO_4$			773K, 1 h	90	3[1]
20MgAAP-573	$4MgO \cdot 13Al_2O_3 \cdot 10AlPO_4$	mix-mulling with nickel carbonate	20	573K, 1 h	20	4[1]
20MgAAP-773	$4MgO \cdot 13Al_2O_3 \cdot 10AlPO_4$			773K, 1 h	100	4[1]
2NB-573	Nb_2O_5	incipient wetness impregnation with nickel nitrate	2	573K, 1 h	100	4[2]
2NB-773	Nb_2O_5			773K, 1 h	100	4[2]
10NB-573	Nb_2O_5	incipient wetness impregnation with nickel nitrate	10	573K, 1 h	100	9[2]
10NB-773	Nb_2O_5			773K, 1 h	100	9[2]
9NS-573	$Nb_2O_5-SiO_2$	incipient wetness impregnation with nickel nitrate	9	573K, 1 h	30	3[2]
9NS-773	$Nb_2O_5-SiO_2$			773K, 1 h	100	6[2]

[1] From magnetic measurements.

[2] From X-ray line broadening measurements using Mo Kα radiation.

earlier, this study focussed on the chemical behavior of these
catalysts in ethane hydrogenolysis and CO hydrogenation. Both
reactions were run in a microreactor operated in a differential
mode at atmospheric pressure. The bracketing technique of Yates et
al. (11) was followed in the case of ethane hydrogenolysis, and
the procedure suggested by Vannice (12) was used in CO hydrogenation. Details of the reactor system and procedure can be found
elsewhere (6). In expressing the activity in terms of turnover
frequency, the number of surface nickel atoms was calculated by
using the average particle size and percent reduction shown in
Table I.

Magnetic Measurements. Magnetization measurements were performed
using a Cahn model 6602-4 Faraday apparatus. Basically two types
of experiments were done. The first type involved measuring the
change of magnetization as a function of temperature at a constant
field strength of 5 kOe, from which magnetization–temperature
behavior and Curie temperature could be obtained. The second type
involved measuring the magnetization as a function of applied
field at room and liquid nitrogen temperature. The particle size
distribution was then obtained by fitting the data to the Langevin
equation (13). A comparison between the saturation magnetization
of the sample, obtained from extrapolating the magnetization–field
curve to infinite field, and that obtained from a calibration
standard (pure bulk nickel) yielded the apparent reducibility of
the catalyst. A more thorough description of data analysis is
given elsewhere (14).

Results

Ethane Hydrogenolysis. Table II summarizes the kinetic results of
ethane hydrogenolysis over nickel catalysts on phosphate (15),
niobia (6), and niobia–silica (9) supports. As a point of reference, Taylor et al. (16) reported an activity range of $10^{-2}-10^{-4}$
molecule/s/Ni at 478 K for Ni/SiO$_2$ catalysts containing 1, 5, and
10 wt % Ni. Of all the samples reduced at 573 K for 1 h, nickel on
MgAAP and NS both have an activity close to that of Ni/SiO$_2$,
indicating a small support effect. On the other hand, the two
niobia–supported samples already show a significant decline in
activity. The activities of all samples decline further after
reduction at 773 K, and the niobia–supported samples remain the
least active.
    All catalysts show roughly a first-order dependence on eth-
ane partial pressure, but the dependence on hydrogen partial
pressure varies. The three most interacting niobia–supported
samples have a reaction order of ∿ -1, whereas the other samples have values between -1.5 and -1.8. Activation energies of all
the samples are close to the value of 170 kJ/mole reported for
Ni/SiO$_2$ (11). The one exception is the 7AAP-773 sample, which has
a somewhat higher value.

Carbon Monoxide Hydrogenation. The activities of these nickel
catalysts in CO hydrogenation are shown in Table III. The phosphate–supported catalysts as a group are about a factor of 5 to 10

Table II. Kinetic Results of Ethane Hydrogenolysis
over Nickel Catalysts

Sample	E_A, kJ/mol	$n^{(4)}$	$m^{(4)}$	Activity at 478 K (molecule/s/atom)
7AAP-773[1]	194	0.9	-1.5	8.0×10^{-7}
20MgAAP-573[1]	160	0.9	-1.8	1.4×10^{-4}
20MgAAP-773[1]	175	0.9	-1.5	2.0×10^{-6}
2NB-573[2]	176	1.1	-1.1	1.1×10^{-6}
2NB-773[2]	180	1.0	-1.2	1.2×10^{-7}
10NB-573[2]	183	0.9	-1.7	4.9×10^{-6}
10NB-773[2]	170	1.0	-0.9	1.4×10^{-7}
9NS-573[3]	179	0.9	-1.7	5.9×10^{-5}
9NS-773[3]	172	0.9	-1.7	4.1×10^{-6}

[1] Taken from reference (15).

[2] Taken from reference (6).

[3] Taken from reference (9).

[4] Exponents in the experimental rate law, $kp_E^n p_H^m$, where p_E and p_H correspond to ethane and hydrogen partial pressure, respectively.

less active than the niobia–supported catalysts. In fact, their activity is comparable to that reported for Ni/SiO$_2$ (17,18). Their selectivity is also similar to that of Ni/SiO$_2$ in that methane is the primary product. Instead of showing the actual product distribution, the selectivity patterns of these catalysts are highlighted in the last two columns in Table III as to whether more higher hydrocarbons are produced than Ni/SiO$_2$, and whether olefinic products are formed. With these comparisons it is obvious that phosphate and niobia behave very differently. It is noteworthy that the niobia–silica support displays all the characteristics of bulk niobia and shows an even higher activity.

Magnetic Results. Table IV summarizes the magnetic results for this series of catalysts. It should be noted that due to the configuration of the Faraday apparatus, the reductions prior to magnetic measurements were done in a static hydrogen atmosphere of about 400 torr. For this reason the apparent reducibility of the catalyst, shown in Table IV as Ni(0)/(Ni total), is not the same as the percent reduction shown in Table I. In general a higher extent of reduction was achieved when the catalyst was reduced in flowing hydrogen under the same temperature and duration. The purpose of the magnetic study was to detect any bulk electronic changes when these catalysts were subjected to a much more severe reduction treatment. Under the more severe condition the difference between static and flow mode of reduction is expected to be small.

Table III. Kinetic Results of CO Hydrogenation
over Nickel Catalysts

Sample	Activity at 548 K $(\times 10^{-2}$ molecule/s atom) N_{CH_4}	N_{CO}	Shift in product distribution?	Olefin formation?
7AAP-773[1]	0.24	0.27	no	no
20MgAAP-573[1]	0.48	0.68	no	no
20MgAAP-773[1]	0.30	0.39	no	no
2NB-573[2]	1.9	4.2	yes	yes
2NB-773[2]	2.6	3.5	yes	yes
10NB-573[2]	2.4	6.3	yes	yes
10NB-773[2]	1.8	5.8	yes	yes
9NS-573[3]	6.3	37	yes	yes
9NS-773[3]	3.8	17	yes	yes

[1] Taken from reference (15).

[2] Taken from reference (6).

[3] Taken from reference (9).

Table IV. Magnetic Results for Nickel Catalysts[1]

Sample	Average particle diameter (nm)[2]	$\dfrac{Ni(0)}{(Ni\ total)}$	Curie temperature (K)
7AAP-773	3	0.2	603
7AAP-873 (16 h)	3	1	< 573
20MgAAP-773	4	0.45	643
20MgAAP-873 (64 h)	4	1	593
10NB-773	7.5	1	643
10NB-873 (16 h)	4.5	0.55	643
9NS-873 (5 h)	4.5	1	633
9NS-873 (16 h)	3.5-4.5	0.8	643

[1] Taken from reference (14).

[2] All values were obtained from magnetic measurements.

The 7AAP sample showed a normally behaved magnetization-temperature curve after a reduction at 773 K for 1 h. However, an anomalous concave downward curve was observed after a rigorous reduction at 773 K for 16 h (Figure 1). The corresponding Curie temperature (<573 K) is significantly less than that of bulk nickel (631 K). For the 20MgAAP sample, a lowering of Curie temperature was observed only after a reduction at 873 K for 64 h.

For the niobia and niobia-silica supported catalysts, the Curie temperature remains virtually unchanged after a reduction at 873 K for 16 h. As shown in Figure 2, a slight decrease in the curvature of the M(T) curve was found for the 1ONB sample reduced at 873 K for 16 to 64 h. This change in curvature corresponds to an apparent decrease in the average nickel particle size. Such an apparent decrease is illustrated by the particle size distribution data shown in Figure 3. Concurrent with this change the amount of Ni(O) in the catalyst decreased to about half of the original value (Table IV). Figure 4 shows qualitatively a similar shift in the particle size distribution for the 9NS sample with increasing severity of reduction. However, the change is slight compared to the 1ONB sample, so is the decrease in the Ni(O) content.

Discussion

A hierarchy recently developed for Ni/Nb$_2$O$_5$ catalysts identifies several characteristics for SMSI behavior [6]. These characteristics include a decrease in ethane hydrogenolysis activity with a change in the rate law and an increase in CO hydrogenation activity with a shift in product distribution. As discussed elsewhere [6], the hierarchy can be understood in terms of an increasing extent of the migration of reduced niobia onto nickel. Results presented previously clearly show that the phosphate-supported catalysts do not display similar characteristics and suggest a different mechanism of interaction between niobia and phosphate. Meriaudeau et al. [4] also concluded that the nature of interaction is different between TiO$_2$ and CeO$_2$. These authors reported that platinum on both supports show some degree of suppression of hydrogen chemisorption, but different behavior is observed for benzene hydrogenation, n-hexane reaction, and CO hydrogenation. Their study and ours demonstrate the importance of using several adsorptive and catalytic probes to fully characterize metal-support interactions. Suppression in hydrogen chemisorption, which is first related to SMSI [1], may be inadequate in defining this complex behavior in some systems.

Magnetic measurements of the niobia- and phosphate-supported catalysts at high reduction temperatures provided a clue as to what the different mechanisms are. The anomalous magnetization-temperature curve observed for the 7AAP sample is similar to that reported by Iida [19] for amorphous NiP alloy films. An alloy formation between nickel and phosphorous would also be consistent with the observed lowering of Curie temperature. It thus seems likely that phosphorous from the support diffuses into the nickel particle leading to the formation of an alloy. Since solid-solid diffusion is a slow process, a long and severe temperature treatment is necessary to induce the observed change in magnetic behavior. The samples used in the kinetic studies are reduced under milder conditions and as such, represent an early stage of this type of interaction. It is thus not surprising that the phosphate-supported catalysts behave very similarly to Ni/SiO$_2$ in CO hydrogenation. Ethane hydrogenolysis, on the other hand, is more sensitive to this interaction as its activity has already

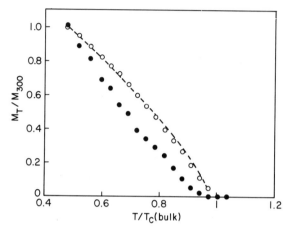

Figure 1.   Magnetization–temperature curves for 7% Ni/AAP.
Reduction treatments were: (O) 773K, 1h; and (●) 873K, 16h.
Dashed line is calculated for a 40 Å diameter   particle with
a Curie temperature of 603 K. (Reproduced with permission from
Ref. 14. Copyright 1985, Academic Press, Inc.)

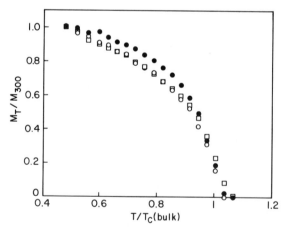

Figure 2.   Magnetization–temperature curves for 10% Ni/Nb$_2$O$_5$.
Reduction treatments were: (●) 773K, 1h; (O) 873K, 16h; and
(□) 873K, 64h. (Reproduced with permission from Ref. 14.
Copyright 1985, Academic Press, Inc.)

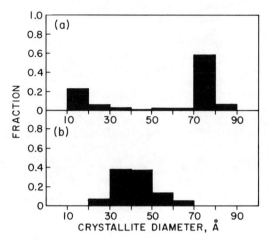

Figure 3.   Particle size distribution data for 10% Ni/Nb$_2$O$_5$. Reduction treatments were: (a) 773 K, 1h; and (b) 873K, 16h. (Reproduced with permission from Ref. 14. Copyright 1985, Academic Press, Inc.)

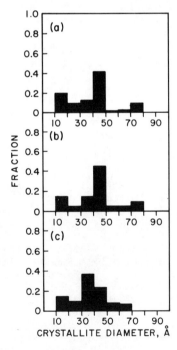

Figure 4.   Particle size distribution data for 9% Ni/Nb$_2$O$_5$– SiO$_2$. Reduction treatments were: (a) 773K, 2h; (b) 873K, 5h; and (c) 873K, 16h.

been lowered. It should be recalled that for Ni/Nb$_2$O$_5$ catalysts, CO hydrogenation is more sensitive than ethane hydrogenolysis as a probe for the interaction (6). Even though it is reasonable to expect such a difference from the dissimilar mechanisms of inter- action for the two systems, future work will be needed to provide a clear physical picture representing the early stage of inter- action in the phosphate–supported catalysts.

Between the two phosphate supports MgAAP is less interactive than AAP under comparable reduction treatment. This difference could be due to the slightly larger particle size of Ni on MgAAP, or a lower availability of phosphorous in the former support. Our present data cannot differentiate these possibilities. The impor- tant point is that a lowering in Curie temperature can also be induced in MgAAP, indicating a similar mechanism of interaction in the phosphate family.

Recently several groups have provided strong evidence for the presence of TiO$_x$ species on nickel from studies on model systems of Ni/TiO$_2$ catalysts (20-22). It is reasonable to expect that oxide migration also occurs in Ni/Nb$_2$O$_5$ in view of the similarity between TiO$_2$ and Nb$_2$O$_5$ (2,6). In fact, the presence of NbO$_x$ species was suggested to explain the observed chemical behav- ior for Ni/Nb$_2$O$_5$ catalysts (6). The following picture then could account for the observed trend in magnetic studies. With the more severe reduction treatment at 873 K for 16 h, it seems plausible that the surface suboxide species increases in concentration and reacts with the nickel metal to form an outer layer of compound. This compound, which is probably nickel niobate, is responsible for the decrease in the overall Ni(0) content. The inner core of nickel still retains its magnetic properties, most notably the Curie temperature, but is of a smaller diameter than the original particle.

Results for the niobia–silica support lend credence to this picture. By limiting the niobia content, a much smaller decrease in either the average particle size or Ni(0) content is found for this sample. The similar trend for the two niobia–containing supports, however, reinforces the conclusions from kinetic results that the same mechanism of interaction prevails in this family of supports.

As noted earlier more severe reduction conditions were used in the magnetic than kinetic studies due to the different scope of these experiments. In order to bridge this gap and fully correlate the results, kinetic measurements were made on a few selected samples reduced at 873 K for 16 h. Nickel supported on AAP, MgAAP, NB, and NS all showed a further decline in ethane hydrogenolysis activity under this severe reduction. The phosphate–supported sam- ples are the least active, which is probably caused by a consider- able extent of alloy formation. A higher activity was found for nickel supported on niobia–silica than on niobia. The lesser degree of interaction for niobia–silica is also evident in the results of CO hydrogenation. As shown in Figure 5, both 10NB and 9NS showed a decrease in activity when the reduction treatment was raised from 773 K for 1 h to 873 K for 16 h. The activity of the 9NS sample is higher at both temperatures. Burch and Flambard (17) also reported a decrease in CO hydrogenation activity with increasing temperature of activation over Ni/TiO$_2$. Thus, it

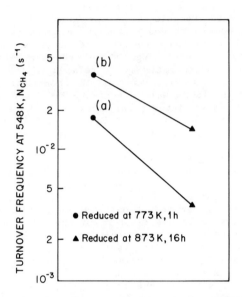

Figure 5. Methanation activity of carbon monoxide hydrogenation over nickel supported on (a) Nb_2O_5 and (b) $Nb_2O_5-SiO_2$ as a function of reduction treatment. Reduction treatments were: (●) 773K, 1h; and (▲) 873K, 16h.

appears generally that for these catalysts the enhancement in
activity first diminishes and then disappears with increasing
severity of reduction. A comparison between catalysts should be
done with this observation in mind.

At present the exact cause for the enhanced CO hydrogenation
activity remains unclear. Several authors have proposed that special active sites exist at the suboxide-metal interface (17,23).
Recently, Raupp and Dumesic (24) suggested that the enhanced
activity of Ni/TiO$_2$ can be understood in terms of more competitive
hydrogen adsorption and that special interfacial sites need not be
invoked. Without differentiating between these models, we simply
suggest that the enhanced activity is related to the presence and
distribution of the suboxide species on the metal surface. The
suboxide species should be well distributed at low concentrations,
thus affording a large interfacial area at its perimeter. With
increasing concentrations these species are likely to aggregate,
resulting in a loss of interfacial area, and eventually lead to
some compound formation at the surface. This qualitative argument
is consistent with the observed activity trend. The use of
niobia-silica presumably slows down the decline in activity by
limiting the concentration of migrating oxide through a low availability of niobia. Consequently the niobia-silica-supported sample
is more active at a given reduction treatment, which would correspond to a more favorable distribution of the suboxide species. It
should be noted that our argument only requires that the activity
is related to the available interfacial area around the perimeter
of the suboxide species. With currently available data it would be
difficult to ascertain whether such a relationship arises from a
creation of special sites at the interface or a difference in the
competitive adsorption of H$_2$ and CO on the metal surface near the
interface.

Summary

Combined kinetic and magnetic studies establish different mechanisms of metal-support interactions for niobia and phosphate supports. The term SMSI, originally coined for TiO$_2$-supported metals,
should perhaps be reserved for systems whose mode of interaction
is similar to that of TiO$_2$. In addition to the suppression in
hydrogen chemisorption and its reversibility with oxidation as
first reported for SMSI behavior, other chemical probes can be
used to further define this particular mode of interaction.

Even for an SMSI oxide, the extent of interaction is dependent on many parameters and a comparison among samples must be
done systematically. Under high reduction temperatures encapsulation of the metal particle leads to a decline in CO hydrogenation
in niobia-containing support. The niobia-silica surface phase
oxide, which shows a similar mechanism of interaction to niobia
but is less interacting, should prove useful as a model system in
future studies.

Acknowledgments

One of us (EIK) would like to thank the National Science Foundation (CPE-8318495) for partial support of this work.

Literature Cited

1.   Tauster, S.J.; Fung, S.C.; Garten, R.L. J. Amer. Chem. Soc. 1978, 100, 170.
2.   Tauster, S.J.; Fung, S.C. J. Catal. 1978, 54, 29.
3.   Adamiec, J.; Wanke, S.E.; Tesche, B.; Klengler, U. In "Metal-Support and Metal-Additive Effects in Catalysis"; B. Imelik et al., Ed.; Elsevier: Amsterdam, 1982, p. 77.
4.   Meriaudeau, R.; Dutel, J.F.; Dufaux, M.; Naccache, C. In "Metal-Support and Metal-Additive Effects in Catalysis"; B. Imelik et al., Ed.; Elsevier: Amsterdam, 1982, p. 95.
5.   Maubert, A.; Martin, G.A.; Prailaud, H.; Turlier, P. React. Kinet. Catal. Lett. 1983, 22, 203.
6.   Ko, E.I.; Hupp, J.M.; Wagner, N.J. J. Catal. 1984, 86, 315.
7.   Marcelin, G.; Vogel, R.F. J. Catal. 1983, 82, 482.
8.   Ko, E.I.; Hupp, J.M.; Rogan, F.H.; Wagner, N.J. J. Catal. 1984, 84, 85.
9.   Ko, E.I.; Bafrali, R.; Nuhfer, N.T.; Wagner, N.J. J. Catal. accepted for publication.
10. Marcelin, G.; Lester, J.E. J. Catal. 1985, 93, 270.
11. Yates, D.J.C.; Taylor, W.F.; Sinfelt, J.H. J. Amer. Chem. Soc. 1964, 86, 2996.
12. Vannice, M.A. J. Catal. 1975, 37, 449.
13. Selwood, P.W. In "Chemisorption and Magnetization"; Academic: New York, 1975; Chapter IV.
14. Marcelin, G.; Ko, E.I.; Lester, J.E. J. Catal. accepted for publication.
15. Ko, E.I.; Marcelin, G. J. Catal. 1985, 93, 201.
16. Taylor, W.F.; Sinfelt, J.H.; Yates, D.J.C. J. Phys. Chem. 1965, 69, 3857.
17. Burch, R.; Flambard, A.R. J. Catal. 1982, 78, 389.
18. Bartholomew, C.H.; Pannell, R.B.; Butler, J.L. J. Catal. 1980, 65, 335.
19. Iida, K. J. Magn. Mater. 1983, 35, 226.
20. Simoen, A.J.; Baker, R.T.K.; Dwyer, D.J.; Lund, C.R.F.; Madon, R.J. J. Catal. 1984, 86, 359.
21. Chung, Y.-W.; Xiong, G.; Kao, C.C. J. Catal. 1984, 85, 237.
22. Raupp, G.B.; Dumesic, J.A. J. Phys. Chem. 1984, 88, 660.
23. Vannice, M.A.; Sudhakar, C. J. Phys. Chem. 1984, 88, 2429.
24. Raupp, G.B.; Dumesic, J.A. Prepr. Div. Petrol. Chem. Amer. Chem. Soc. 1985, 30(1), 137.

RECEIVED September 12, 1985

14

Intermetallic Compounds as Models for Materials Formed at the Metal Crystallite–Oxide Support Interface

Ralph G. Nuzzo and Lawrence H. Dubois

AT&T Bell Laboratories, Murray Hill, NJ 07974

This paper discusses transition metal intermetallic compounds, in the context of the reactivity and physical properties expected for materials produced via solid-solid reactions at the metal catalyst oxide-support interface. It is shown that several observable and proposed features of the so called Strong Metal-Support Interaction (SMSI) — chemisorption activity, phase segregation, and encapsulation — follow naturally from the chemistry of these materials. Both literature precedent and experimental data are presented to support the close relationship suggested above.

Are intermetallic compounds and their associated properties a reasonable model for the Strong Metal-Support Interaction (SMSI)? Before addressing this question, it would serve well to restate and review briefly the most frequently cited explanations of the origin of this effect [1,2]: (1) Direct electron transfer from the support to the metal catalyst; (2) Schottky barrier formation at the metal crystallite-oxide support interface; (3) Changes in crystallite size, structure, and morphology; (4) Encapsulation; (5) Alloy and/or intermetallic compound formation. The first two proposals, which we reasonably classify as being electronic theories, have been the subject of considerable criticism [2]. Indeed, the physics related to the latter of these first two proposals, and metal-semiconductor interfaces in general [3], places almost intolerable constraints on the applicability of this proposed origin of the SMSI effect. It is not our intention or inclination to recite or develop this type of critique. Rather, we would like to describe in general terms the results emerging from our studies on the properties of silicon based intermetallic compounds and to suggest that similar materials and processes might serve centrally in the last three mentioned explanations of the SMSI effect. Indeed, this notion is not completely new to us as Tauster et al., in their original paper [4], showed that the formation of Pt_3Ti (produced via the high temperature reduction of platinum on TiO_2) is thermodynamically feasible. By way of reference, we would direct the reader to the primary articles where much of the following information is discussed in explicit detail [5-11].

Three main properties have come to characterize the SMSI effect [1]. The first of these is a diminished activity toward the chemisorption of H_2 and CO induced by the high temperature reduction of a supported metal catalyst (low temperature reductions are ineffective). The second is a significant alteration of the catalytic

0097–6156/86/0298–0136$06.00/0
© 1986 American Chemical Society

activity and/or selectivity of the metal. Finally, the SMSI state can be reversed, in that these effects are eliminated by a subsequent oxidation at a moderate temperature followed by a low temperature reduction. We first are led to ask, then, do intermetallic compounds exhibit "altered" and diminished chemisorption activity?

The answer to the above is a clear and emphatic yes as first shown by ourselves on the (111) and (100) single crystal surfaces of $NiSi_2$ (5) and more recently by Ross and coworkers on both polycrystalline and single crystalline Pt_3Ti (12,13). We note that, although most studies to date have dealt with zero-valent materials, partially oxidized species may show similar effects. In the case of metal crystallites supported on reducible oxide substrates (14), these latter materials may, in fact, be more relevant. Despite this, the properties of zero valent intermetallic surfaces can provide useful insights as we shall show below.

Chemisorption Studies

Data for the chemisorption of small molecules on $NiSi_2$(111) (5), Pt_3Ti (12,13), Ni_3Ti (15), Ni, Si, Ti, and Pt surfaces are given in Table I. Examination of the above yields two striking contrasts.

Table I

Small Molecule Chemisorption on Intermetallic Compound Surfaces

Surface	O_2	H_2	Adsorbate CO	CO_2
$NiSi_2$	O^a	$NR^{b,c}$	$C+O^d$	$O+CO_{(g)}$
Ni	O	H	CO	NR
Si	O	NR^c	NR	—
Pt_3Ti	O	NR	CO^e	—
Pt	O	H	CO	NR
Ti	O	H	C+O	O+CO
Ni_3Ti	O	H	C+O, CO	—

a. Low exposures, chemisorbed oxygen; high exposures, SiO_2 formation.

b. NR = no reaction.

c. H atoms will chemisorb.

d. Low sticking probability for the molecular species.

e. Amount of CO adsorbed is significantly less than that observed on pure Pt.

First, $NiSi_2$(111) exhibits little or no activity toward the dissociative chemisorption of molecular hydrogen. Pt_3Ti also shows a lack of reactivity toward H_2 (this dissociation is facile on pure nickel and platinum surfaces). Further, our studies show that, on $NiSi_2$(111), this effect is kinetic in origin as hydrogen atoms readily and strongly chemisorb on this surface (5). Similar kinetic limitations to hydrogen dissociation also have been observed on several TiO_2 supported Group VIII metal surfaces (16).

Second, CO strongly chemisorbs as the molecular species on nickel and platinum metal surfaces under the conditions given in Table I while, on $NiSi_2(111)$, the sticking probability is extremely low. Once bound, however, dissociation to surface bound carbon and oxygen occurs in a facile process. Our subsequent studies have shown that these properties are not singular to the $NiSi_2(111)$ surface (i.e. a surface in which all Ni-Ni and Si-Si bonding is excluded) as both different nickel silicide stoichiometries and morphologies have shown similar perturbations of chemisorption activity. Facile CO dissociation was also seen on Ni_3Ti (15). In sharp contrast, Pt_3Ti shows both a decrease in total CO uptake and a shift to lower binding energy compared to pure platinum; no dissociation was detected (12). It is thus clear that intermetallic compound formation has not made the metal centers in these materials uniformly either more or less reactive. Bardi et al. have made similar observations and ascribed them to alterations of electronic structure arising predominately through a ligand effect (12). Such notions are hard to generalize, however. For example, the electronic structure of $NiSi_2(111)$ has been likened unto that of a compound noble metal (17), a description which seems at odds with its high reactivity towards CO.

Catalytic Activity

Studies of the catalytic chemistry of bulk metal silicides are difficult to perform due to their facile oxidation upon exposure to air (7,18,19). It was necessary, as a result, to prepare catalysts in situ (i.e., in the reactor). There are reports in the literature that supported intermetallic compounds can be prepared by heating nickel on silica (an SMSI example using a highly redox resistant support) catalysts to 900°C in flowing H_2 (20,21). In fact, the semiconductor literature is replete with studies showing that intermetallic compound thin films can be formed via the high temperature deposition of metal overlayers on oxidized silicon single crystal substrates (22,23). In the case of the catalytic studies, involving metals supported on silicon oxide carriers, the new materials were characterized by a significant decrease in hydrogen uptake activity as well as by a decrease in saturation magnetization (20). Further studies of such reactions in our laboratory indicated that both the metal and the support were extensively sintered. These samples displayed low CO and H_2 chemisorption activity as well as a slightly altered activity and selectivity in the competitive dehydrogenation/hydrogenolysis of cyclohexane (6).

   In order to form well characterized intermetallic materials under more mild conditions, we have prepared supported silicon based intermetallic compounds by the metal surface catalyzed decomposition of organosilanes (6,7). The decomposition of such reagents as SiH_4 or $(CH_3)_6Si_2$ (the latter in the presence of H_2) on nickel surfaces cleanly yields intermetallic compounds whose structure and stoichiometry depend on such factors as substrate temperature, total gas exposure, and reaction time. The generality of this procedure is indicated on the periodic table given in Figure 1 which details the metals we have examined and the conditions necessary to effect their modification (7). XPS and Auger studies indicate that the procedures of thin film interdiffusion (a standard technique for forming intermetallic compounds) and surface mediated decomposition yield materials of nearly identical stoichiometry when reactions are run under comparable conditions (7). As expected, we observed low H_2 chemisorption activity on both supported and unsupported intermetallic materials formed by this latter preparative procedure. CO chemisorption activity was not studied for the reasons discussed below.

The trends observed in the chemisorption activity of the nickel silicides (see above) are also manifested in their catalytic activity (6,8). This is most clearly shown by the data given in Figure 2. As is indicated, the catalytic reformation of cyclohexane over a supported nickel silicide catalyst exhibits both different product balances and power rate dependences than that which obtains for pure nickel. We believe that the negligible H_2 partial pressure dependence of this reaction is the result of a large kinetic barrier to hydrogen dissociation on the nickel silicide surface. Several hydrogenation and isomerization studies on Pd-Si glasses have also shown significant changes in catalytic activity and selectivity when compared to reactions taking place on pure palladium (24,25). Thus, in terms of important chemical reactivity patterns, intermetallic compounds can exhibit properties akin to those seen in the SMSI effect.

Although not expressly stated as such, extensions of the data given above also begin to address the question of reversibility. As we will show, there exists a strong relationship between the ideas which follow and the proposed importance of encapsulation and morphological changes in SMSI.

Surface Oxidation

Table I shows us that O_2 dissociates readily on $NiSi_2(111)$. Examination of high resolution electron energy loss (EELS) spectra leads to the suggestion that this adsorption results in the initial coordination of oxygen atoms at bridged nickel-silicon sites (5). Aging the sample, or simply warming it to ambient temperature, results in the observation of a very different type of oxygen centered bonding. New modes appear which indicate a significant restructuring of the surface to generate silicon based bonding in Si-O-Si linkages, that is, the preferential segregation of a silicon oxide (5). This effect is powerfully demonstrated by the XPS data shown in Figure 3. The upper trace shows core levels of Si (2s and unresolved 2p doublet) and Ni (3s) as they appear in a Ni_2Si thin film grown by the surface mediated decomposition of SiH_4 on a nickel foil at 300°C. Exposure of this material to air at 25°C yielded the surface characterized by the spectrum in the middle trace in Figure 3. Careful examination of these and other core levels show that oxidation, under these conditions, results in the preferential segregation of oxidized silicon (7,9,11). Further, the metal silicide *underlayer* is *passivated* by this thin non-native oxide. The results of Bardi et al. show that this is not an exclusive property of silicon based intermetallic compounds; both Pt_3Ti (12,13) and Pt-Zr (26) thin films show similar preferential segregation of the thermodynamically more stable oxide. In addition, materials such as Ti_2Ni, $TiCu_3$, and iron-titanium alloys (27) as well as $LaNi_5$ (28) require pre-treatment at elevated H_2 pressures to remove an encapsulating surface oxide before significant hydrogen absorption can take place.

We, and others, have also found that these same oxidations are very hard to prevent even under strongly "reducing" or UHV conditions (5,29). For example, Table I shows that even CO and CO_2 are sufficiently strong "oxidants" to effect these reactions while other observations we have made suggest that low level water impurities in hydrogen streams can result in the oxidation of the intermetallic with the concomitant encapsulation and passivation of the metal. The consequences of oxidation at high temperatures have been shown in convincing detail by Wallace (30), Hercules (19), and their coworkers. All elements of the intermetallic are oxidized and the morphology and chemical state of the metal particles, as a result, are extensively altered. A subsequent reduction at low temperature reduces only the group VIII

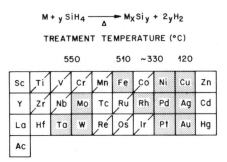

Figure 1. Periodic table showing intermetallic compound formation from the reaction of clean metal surfaces with ~5000L of SiH$_4$. Temperature required to grow thin films are indicated above the table. Solid areas indicate metals studied in our laboratory, while the cross-hatched areas refer to elements which should be employable based on thin film interdiffusion precedents.

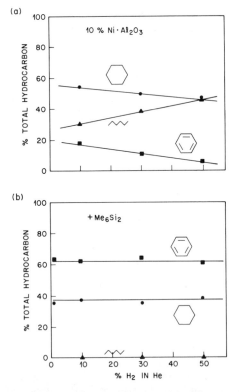

Figure 2. Product distribution for cyclohexane conversion to either benzene or hydrogenolysis (>90% *n*-hexane) products over (a) pure nickel on alumina and (b) the same catalyst after treatment with hexamethyldisilane in H$_2$. Reaction conditions are discussed in references (6) and (8).

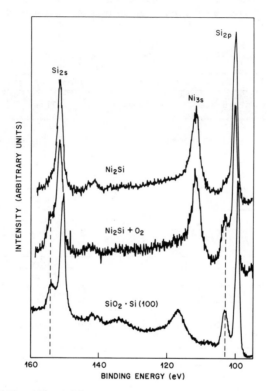

Figure 3. XPS spectra of (a) a Ni_2Si thin film formed via the reaction of a clean Ni foil with SiH_4 at 300°C; (b) same foil after exposure to air for 5 minutes at 30°C; (c) silicon (100) single crystal with a native oxide overlayer.

component (that is to say, the "SMSI effect" is reversed and a novel oxide supported metal catalyst is formed (30)). A *schematic* view of the entire process — intermetallic compound formation at elevated temperatures, oxidation, and subsequent reduction — is shown in Figure 4. The general notion presented here, then, is that intermetallics can provide an effective means of transport of a support metal ion to the surface of a catalyst.

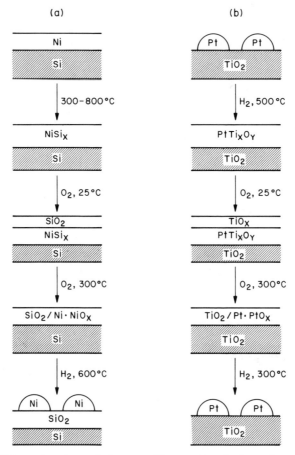

Figure 4. *Schematic* view of intermetallic compound formation, oxidation (encapsulation), and subsequent reduction of (a) Ni on Si and (b) Pt on TiO_2. We note that the oxidation states of Pt and Ti in $PtTi_xO_y$ are not known (i.e., y may be 0); this figure is intended to indicate qualitative phase formation characteristics only.

General Observations

What, then, need pertain in order for intermetallic compounds to help explain what has been called SMSI? First, and most importantly, they must form as a result of the high temperature reduction of the metal catalyst (typically Group VIII) on an appropriate support (generally TiO_2). We conclude that there is ample precedent for this type of reactivity in both the chemical (4,20,21) and thin film technology (22,23) literature. It should be noted, in this regard, that the SMSI effect is most frequently, but not exclusively, observed on metal oxide supports which are easily reduced (14). Second, these materials, *or the products derived from them*, must exhibit the observable properties characterizing SMSI. Our belief is that the studies reviewed above, as well as those of others not directly cited herein (2), allow strong inferences to be made about the relationship between the properties of intermetallic compounds, and their derivatives, and the mechanism(s) and characteristics of the SMSI effect. Indeed, it may prove that SMSI is not a single effect (31) under all conditions of practice. This is most clearly suggested by the multi-phase material whose core level spectrum is shown in the middle trace in Figure 3. This material, comprised of layered pure metal, intermetallic, and passivating/inert oxide domains (Figure 4), shows how a complex morphology, characterized by abnormal chemisorption activity, is easily obtained as a result of intermetallic compound formation followed by a subsequent selective segregation of an "inert encapsulating oxide."

Literature Cited

(1)   Tauster, S. J.; Fung, S. C.; Baker, R. T. K.; Horsley, J. A. *Science* **1981**, *211*, 1121-1125.

(2)   Specific details and references can be found in the accompanying papers in this symposium as well as in the volume *Stud. Surf. Sci. Catal.*, Imelik, B.; Naccache, C.; Coudurier, G.; Praliaud, H.; Meriaudeau, P.; Gallezot, P.; Martin, G. A.; Verdine, J. C., Eds.; Vol 11, 1982.

(3)   Sze, S. M. "Physics of Semiconductor Devices"; John Wiley and Sons: New York, 1981.

(4)   Tauster, S. J.; Fung, S. C.; Garten, R. L. *J. Am. Chem. Soc.* **1978**, *100*, 170-175.

(5)   Dubois, L. H.; Nuzzo, R. G. *J. Am. Chem. Soc.* **1983**, *105*, 365-369.

(6)   Nuzzo, R. G.; Dubois, L. H.; Bowles, N. E.; Trecoske, M. A. *J. Catal.* **1984**, *85*, 267-271.

(7)   Dubois, L. H.; Nuzzo, R. G. *J. Vac. Sci. Technol.* **1984**, *A2*, 441-445.

(8)   Nuzzo, R. G.; Dubois, L. H. *Appl. Surf. Sci.* **1984**, *19*, 407-413.

(9)   Nuzzo, R. G.; Dubois, L. H. *Surf. Sci.* **1985**, *149*, 119-132.

(10) Dubois, L. H.; Nuzzo, R. G. *Surf. Sci.* **1985**, *149*, 133-145.

(11) Dubois, L. H.; Nuzzo, R. G. *Langmuir* in press.

(12) Bardi, U.; Somorjai, G.A.; Ross, P. N. *J. Catal.* **1984**, *85*, 272-276.

(13) Bardi, U.; Ross, P. N. *J. Vac. Sci. Technol.* **1984**, *A2*, 1461-1470.

(14) Tauster, S. J.; Fung, S. C. *J. Catal.* **1978**, *55*, 29-35.

(15) Fischer, T. E.; Kelemen, S. R.; Polizzotti, R. S. *J. Catal.* **1981**, *69*, 345-358.

(16) Jiang, X.; Hayden, T. F.; Dumesic, J. A. *J. Catal.* **1983**, *83*, 168-181.

(17) Chabal, Y. J.; Haman, D. R.; Rowe, J. E.; Schlüter, M. *Phys. Rev.* **1982**, *B25*, 7598-7602.

(18) See for example Valeri, S.; Pennino, V. D.; Lomellini, P.; Sassaroli, P. *Surf. Sci.* **1984**, *145*, 371-389 and references cited therein.

(19) Honalla, M.; Dang, T. A.; Kibby, C. L.; Petrakis, L.; Hercules, D. M. *Appl. Surf. Sci.* **1984**, *19*, 414-429.

(20) Praliaud, H.; Martin, G. A. *J. Catal.* **1981**, *72*, 394-396.
(21) Similar results have been reported for platinum on alumina catalysts. See den Otter, G. J.; Dautzenberg, F. M. *J. Catal.* **1978**, *53*, 116-125.
(22) Tu, K. N.; Mayer, J. W. in "Thin Films – Interdiffusion and Reactions"; Poate, J. M.; Tu, K. N.; Mayer, J. W., Eds.; J. Wiley: New York, 1978; Chapt. 10 and references cited therein.
(23) Ottaviani, G. *J. Vac. Sci. Technol.* **1979**, *16*, 1112-1119 and references cited therein.
(24) Brower, W. E., Jr.; Matyjaszczyk, M. S.; Pettit, T. L.; Smith, G. V. *Nature* **1983**, *301*, 497-499.
(25) Smith, G. V.; Zahraaj, O.; Molnar, A.; Khan, M. M.; Rihter, B.; Brower, W. E. *J. Catal.* **1983**, *83*, 238-241.
(26) Bardi, U.; Ross, P. N.; Somorjai, G. A. *J. Vac. Sci. Technol.* **1984**, *A2*, 40-49.
(27) Padurets, L. N.; Sokolova, E. I.; Kost, M. E. *Russ J. Inorg. Chem.* **1982**, *27*, 763-765.
(28) Siegmann, H. C.; Schlapbach, L.; Brundle, C. R. *Phys. Rev. Lett.* **1978**, *40*, 972-975.
(29) Imamura, H.; Wallace, W. E. *J. Phys. Chem.* **1979**, *83*, 2009-2012.
(30) Imamura, H.; Wallace, W. E. *J. Phys. Chem.* **1979**, *83*, 3261-3264.
(31) This notion has been broadly developed by others. See Belton, D. N.; Sun, Y.-M.; White, J. M. *J. Phys. Chem.* **1984**, *88*, 5172-5176 and references cited therein.

RECEIVED September 17, 1985

Effect of H$_2$ Treatment on the Catalytic Activity of Pt–SiO$_2$ Catalysts

H. Zuegg and R. Kramer

Institut für Physikalische Chemie, Universität Innsbruck, A-6020 Innsbruck, Austria

The deactivation of platinum catalysts for hydrocarbon reactions caused by high temperature reduction (HTR) has been observed by many authors (for review, see 1). When using titania as support the extent of deactivation is especially strong, but also with "nonreducible" supports like silica or alumina similar effects have been reported (2, 3). The aim of this work was to investigate the changes in activity of Pt/SiO$_2$ model catalysts caused by HTR. Special interest has been paid to the activity change of the Pt/SiO$_2$ phase boundary in relation to the bulk platinum surface, as was for instance observed recently by Resasco and Haller (4) with Rh/TiO$_2$. We have shown earlier (5, 6) that the hydrogenolysis of methylcyclopentane (MCP) on platinum proceeds via two different pathways, (i) occurring at the "bulk" platinum surface and yielding exclusively 2-methylpentane (2-MP) and 3-methylpentane (3-MP), and (ii) occurring at the phase boundary Pt-support, where nearly statistical ringopening leads to the formation of 2-MP, 3-MP and n-hexane (n-H). The selectivity for n-hexane formation in this reaction was therefore taken to indicate possible activity changes of the phase boundary in relation to the Pt-surface. Parallel to the MCP hydrogenolysis the hydrogenation of benzene was chosen for examining in addition the effect of HTR on a structure insensitive reaction.

Experimental

The model catalysts were prepared by HV deposition of an amorphous SiO$_2$ film (by evaporation of SiO in 10^{-2} Pa of oxygen), onto which Pt was deposited by high vacuum evaporation, as described earlier (7). In order to get catalysts of different dispersion, the mean thickness of deposited Pt was varied between 0.1 and 1 nm. By TEM inspection of catalyst specimens the particle density and the particle size distribution were obtained, from which data the platinum surface area and the dispersion were calculated. Additionally a conventional 6,3 % Pt/SiO$_2$ catalyst (EUROPT-1, d̄ = 1.7 nm) was used in the experiments.
The catalytic reactions were carried out in an all-glass recirculation apparatus providing long reaction times. For all reactions the conversion-time behavior was measured up to a reaction

0097–6156/86/0298–0145$06.00/0

time of one hour. The partial pressures of MCP and benzene were
1 KPa and 3 KPa respectively, hydrogen was admitted to a total
pressure of 1 atmosphere. The reaction temperatures were 523 K for
MCP hydrogenolysis and 323 K for benzene hydrogenation. Before each
experiment the catalysts were pretreated by heating in oxygen at
673 K. After cooling down to room temperature in flowing helium
the catalysts were reduced by heating up in hydrogen (a) to 523 K for
low temperature reduction (LTR) and (b) to 673 K for high temperature
reduction (HTR). The LTR was continued for at least one hour at
523 K, in the HTR the time of treatment was varied in order to
investigate the extent of deactivation as a function of HTR treatment
time.

Results

A. Hydrogenolysis of Methylcyclopentane.   After LTR of the
catalysts the product distribution of the MCP hydrogenolysis shows
the expected dependence on particle size.(Table 1). The activities
of all LTR-catalysts agree within experimental error, leading to an
approximate turnover number of 40 h^{-1}, independent of the metal
dispersion. The conversion increases nearly linearly with reaction
time, indicating that no significant deactivation occurs during
reaction. The HTR causes generally both a decrease in the
selectivity for n-hexane and a loss of catalytic activity, but
these changes exhibit a different time dependence. After HTR for only
15 minutes the full shift in selectivity is reached, while the
activity has changed only moderately. After HTR for several hours the
remaining activity reaches a stationary level of about 10 to 40 % of
the initial value (Fig. 1). No general dependence of the deactivation
on dispersion could be established. However, with the EUROPT-1
catalyst only a change in the reaction selectivity but no significant
loss in the activity was observed. The changes of the catalytic
behavior turned out to be reversible and initial selectivity could
be reestablished by treatment in oxygen, but again selectivity was
changed more easily than the activity. While treatment of the
catalysts with oxygen even at room temperature gave the initial
selectivity, the activity was only partially restored after room
temperature oxygen treatment. Further increase of activity was observed by oxygen treatment at successively higher temperatures
(Fig. 2). Only when treated at 673 K in oxygen did the catalysts
regain their full initial activity.

B. Hydrogenation of benzene.   After LTR the activities of the
catalysts agree within experimental error resulting in a turnover
number of approximately 300 h^{-1}. Again, after HTR the activity of the
model catalysts was decreased, while the EUROPT-1 catalyst retained
the initial activity. With the model catalysts the extent of
deactivation by HTR is smaller for benzene hydrogenation than for MCP
hydrogenolysis. However, oxygen treatment at room temperature has a
smaller effect on benzene hydrogenation activity than on the
activity for MCP hydrogenolysis. For both reactions the same fraction
of initial activity is regained after this treatment and the further
recovery of activity due to oxygen treatment at successively higher
temperatures proceeds in a similar way for both reactions.

Table I.   Effect of HTR on catalytic behavior.

Catalyst		Selectivity change		Activity change	
Pt mean thickness (nm)	Dispersion %	% n-hexane in MCP hydrogenolysis after LTR	after HTR	Activity ratio HTR/LTR for MCP hydrogenolysis	for benzene hydrogenation
1.0	7	3	3	0.3	0.7
0.52	17	6	8	0.3	0.8
0.3	20	12	11	0.1	0.8
0.23	30	28	23	0.4	1.0
0.1	40	37	32	0.15	0.7
EUROPT-1	60	40	33	1.0	1.0

Discussion

The results obtained with the MCP hydrogenolysis indicate that HTR causes two different effects on the catalytic behavior of the Pt/SiO$_2$ model catalysts: (i) a fast change of the selectivity to less n-hexane formation followed by (ii) a slow decrease of activity.

As was mentioned above the nonselective pathway of MCP hydrogenolysis (formation of n-hexane additional to 2-MP and 3-MP) takes place at the phase boundary platinum-support (6). From the change of selectivity towards less n-H formation it is concluded that HTR causes a preferential inhibition of catalytic ensembles composed by platinum and support sites. This process occurs rather fast but requires reduction temperatures higher than 600 K. In earlier work (8) we were able to show that at these temperatures atomic hydrogen can be adsorbed at the support surface via hydrogen spillover. The support surface next to the platinum should quickly be saturated by this spilled-over hydrogen resulting in a partial reduction of the silica surface. This reduction of the support surface is assumed to be responsible for the change of selectivity by inhibition of the catalytic sites situated at the phase boundary platinum-silica. The inhibition of these "adlineation ensembles" may occur either by poisoning of support sites by atomic hydrogen or by formation of a bond between platinum and silica, as has been proposed recently by Frennet and Wells (9), resulting in a changed electronic structure of the platinum atoms adjacent to the support. A localized charge transfer has been proposed recently by Resasco and Haller (4) to account for the activity loss of the Rh/TiO$_2$ interface.

The fast attainment of the initial selectivity by an oxygen treatment at room temperature is also understandable on the basis of this mechanism, since atomic hydrogen adsorbed at support sites adjacent to the platinum should easily be oxidized even at room temperature.

The overall deactivation due to the HTR is reversed by oxygen treatment at 673 K. Hence sintering of the platinum particles is not responsible for the deactivation since redispersion is not likely to

occur in oxygen at 673 K. Furthermore the deactivation is only
observed with the model catalysts, while the activity of the EUROPT-1
catalyst is hardly affected by HTR. The support of the model catalysts is prepared differently than that of the EUROPT-1 and we
therefore conclude that the supporting material is involved in the
deactivation process. This result rules out the proposal during HTR
inhibit platinum sites, since in this case the effect of HTR should
not depend on the support.

The HTR affects more strongly the activity for MCP-hydrogenolysis
than that for benzene hydrogenation. However, as is seen in Figure 2,
even an oxygen treatment at room temperature results in a recovery of
activity for MCP hydrogenolysis to reach the same fraction of initial
activity as for benzene hydrogenation. The stronger deactivation vs
MCP hydrogenolysis due to HTR can be related to the change in
selectivity, i.e. both are caused by HTR and are reversed by oxygen
treatment at room temperature. Therefore we conclude that the
partial reduction of the support also influences the activity of the
platinum, at least for hydrogenolysis of MCP. Oxygen treatment at
room temperature most likely results in the reoxidation of the
support surface thereby reversing the special deactivation and the
change of selectivity However the full activity is not regained by
this oxygen treatment at room temperature.

As the course of reactivation by further oxygen treatments agrees
for both reactions, we assume that part of the platinum surface might
be inhibited geometrically by support material, a mechanism well
established for titania supported catalysts exhibiting SMSI behavior
(10). This inhibition of the platinum surface could occur either by
surface migration of silica due to wetting conditions in the reducing
ambient causing the formation of a silica skin or by diffusion of
molecular SiO_x species onto the platinum surface.

The formation of a silica skin has been already proposed by
Schuit et al. (11) for explaining deactivation effects in Ni/SiO_2
catalysts. From electron microscopic observations the beginning of
skin formation in the Pt/SiO_2 system at high temperatures was
deduced (12).

On the other hand, the segregation of silica on platinum surfaces
has been reported to start at about 600 K by bulk diffusion through
platinum (13). Van Langeveld et al. (14) have shown that the
structure of the substrate (quartz or Pyrex) determines, whether
SiO_x species can migrate through evaporated films of platinum. Hence,
the fact that EUROPT-1 does not suffer from this deactivation could
be due to the different preparation procedures of the silica. The
support for the model catalysts, which is prepared by deposition of
SiO in 10^{-2} Pa of oxygen, may have retained some structural "memory"
on the formation from SiO and may therefore be reduced more easily
than the conventional silica of the EUROPT-1 catalyst. The islands
formed by segregation of silicon from bulk platinum are reported to
be easily oxidized to SiO_2 at temperatures lower than 400 K, when
they are exposed to atomic oxygen (15). Thus assuming molecularly
dispersed silicon species to be formed during HTR, these species
should be readily oxidized to SiO_2 even at room temperature, since
atomic oxygen necessary for this oxidation is certainly provided by
dissociatively adsorbed oxygen on platinum.

While oxygen treatment at room temperature is assumed to be
sufficient to cause reoxidation of either the silica skin or the

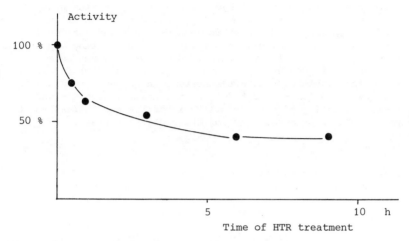

Figure 1. Loss of activity towards MCP hydrogenolysis as a function of HTR treatment time.

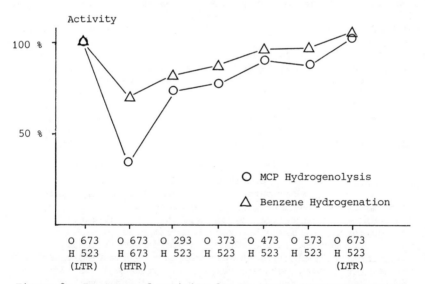

Figure 2. Recovery of activity due to oxygen treatments at successively higher temperatures.

silicon islands formed during HTR, higher temperatures are obviously
necessary for removing the silica from the platinum surface thereby
restoring the activity for MCP hydrogenolysis and benzene hydrogenation in a similar way. The mobility in the silica/platinum system
also under oxidizing conditions has been demonstrated earlier ($\underline{5}$) by
an experiment, using a Pt/SiO$_2$ model catalyst, exhibiting a platinum
particle size of about 10 nm. This catalyst was covered by a 20 nm
thick layer of SiO$_2$ by HV deposition, resulting in a total loss of
catalytic activity. However, after heating in oxygen at 673 K
followed by reduction at 523 K 26 % of the initial activity was
regained, indicating that the mobility is sufficient to lead to the
reexposure of a substantial part of the platinum surface.

Conclusion

(i) High temperature reduction of platinum /silica model catalysts
causes two different effects (a) change in product distribution
connected with a special deactivation of MCP hydrogenolysis and (b)
slow overall deactivation for both MCP hydrogenolysis and benzene
hydrogenation.
(ii) The special deactivation of MCP hydrogenolysis and the change in
product distribution is assumed to be due to a partial reduction of
the support next to the platinum via hydrogen spillover causing a
deactivation of catalytic ensembles at the phase boundary. The deactivation of the platinum support ensembles may be caused either by
inhibition of support sites by atomic hydrogen or by formation of a
Pt-Si bond leading to a disturbed electronic structure of platinum
adjacent to the support.
(iii) The overall deactivation for MCP hydrogenolysis and benzene
hydrogenation is most likely due to coverage of platinum by silica.
This coverage can occur either by silica skin formation due to wetting
conditions in a reducing atmosphere or by diffusion of molecularly
dispersed SiO$_x$ species onto the platinum particles.
(iv) Recovery of initial activity is attained by oxygen treatment at
673 K. It is assumed that this treatment causes the removal of
the silica skin or the diffusion of the SiO$_2$ islands back to the
support.

The effect of hydrogen pretreatment temperature on the catalytic
behavior of Pt/SiO$_2$ model catalysts has been studied for hydrogenolysis of methylcyclopentane (MCP) and hydrogenation of benzene. For
benzene hydrogenation the catalysts exhibit only a slightly lower
activity after high temperature reduction compared to that after low
temperature reduction. In the MCP hydrogenolysis both a loss of
activity and a change of product distribution is observed after high
temperature reduction. However, the initial product distribution of
MCP hydrogenolysis is reestablished after oxygen treatment at room
temperature, and full initial activity for both reactions is regained after oxygen treatment at 673 K.
On the basis of these results the following mechanisms are
deduced: (i) Partial reduction of the support adjacent to the platinum via hydrogen spillover, resulting in a special deactivation for
MCP hydrogenolysis and in a change of product distribution. This

reduction is reversed by oxygen treatment at room temperature.
(ii) The overall deactivation for both reactions is assumed to be
due to platinum covered by supporting material. The silica cover
layer is assumed to be formed either by formation of a silica skin
due to wetting conditions in the reducing atmosphere, or by bulk or
surface diffusion of SiO_x species onto the platinum particles. By
oxygen treatment at 673 Ř the silica cover layer is removed and the
full platinum surface is again exposed to the gas phase.

References

1. Bond, G.C. and Burch, R., in "Catalysis", ed. G.C. Bond and
   G. Webb, The Royal Society of Chemistry, London, 1983,Vol. 6,p. 27
2. Menon, P.G. and Froment, G.F., J. Catal. <u>59</u> , 138 (1979)
3. den Otter, G.J. and Dautzenberg, F.M., J. Catal. <u>53</u>, 116 (1978)
4. Resasco, D.E. and Haller G.L., J. Catal. <u>82</u>, 279 (1983)
5. Kramer, R. and Zuegg, H., J. Catal. <u>80</u>, 446 (1983)
6. Kramer, R. and Zuegg, H., in Proceedings of the 8th Intern.
   Congress on Catalysis, Berlin 1984, Vol. 5, p. 275
7. Kramer, R. and Zuegg, H., J. Catal. <u>85</u>, 530 (1984)
8. Kramer, R. and Andre, M., J. Catal. <u>58</u>, 287 (1979)
9. Frennet, A. and Wells, P.B., Applied Catal., in press
10. Anderson, J.B.F., Bracey, J.D., Burch, R. and Flambard, A.R.,
    in Proceedings of the 8th Intern. Congress on Catalysis, Berlin
    1984, Vol. 5, p. 111
11. Schuit, G.C., and van Reijen, L.L. in "Advances in Catalysis"
    Vol. 10, p. 242. Academic Press, New York/London 1958
12. Powell, B.R., and Whittington, S.E., J. Catal., <u>81</u>, 382 (1983)
13. Niehus, H. and Comsa, G., Surf. Sci. 102 (1981) <u>L</u> 14
14. van Langeveld, A.D., Nieuwenhuys, B.E., and Ponec, V.,
    Thin Solid Films, 105 (1983) 9
15. Bonzel, H.P., Franken, A.M. and Pirug, G., Surf Sci., 104
    (1981) 625

RECEIVED September 17, 1985

16

Interactions and Surface Phenomena in Supported Metal Catalysts

E. Ruckenstein

Department of Chemical Engineering, State University of New York at Buffalo, Amherst, NY 14260

Transmission Electron Microscopy was used to investigate the behaviour of Fe/alumina catalysts in various
environments (such as H_2 and O_2). The results indicate that depending on the environment, the crystallites can extend, spread or contract upon the substrate
and also acquire toroidal shapes or fracture into
smaller units. At high temperatures and in an O_2
atmosphere, films spread out from the crystallites
and a (multilayer) contiguous film coexists with the
crystallites. These wetting and spreading phenomena
are a result of the interactions between the metal,
substrate and atmosphere. The strong chemical interactions between the substrate and the compounds
formed between the active metal and the chemical
atmosphere can enormously decrease the interfacial
free energy between the crystallites and substrate.
This then leads to a rapid spreading of the crystallites, to torus formation and also fragmentation of
the crystallites. New mechanisms of sintering
based on wetting and spreading have been suggested
to explain some experimental observations.
    Considerations based on spreading are used to
explain the room temperature suppression of CO and H_2
chemisorption upon any of the Group VIII metals supported on TiO_2 when the supported metal is prereduced
in H_2 at 500°C (The Tauster effect). From the condition that a layer of oxide should spread over the
metal, one concludes that only those combinations
of oxides and metals for which the interaction energy
per unit interfacial area is greater than twice the
surface free energy of the oxide can manifest the
Tauster effect. A thin film thermodynamic treatment
is employed in which the free energy of formation of
the oxide film is expressed in terms of the film
thickness. The minimization of this free energy
with respect to the film thickness provides an ex-

0097-6156/86/0298-0152$06.00/0
© 1986 American Chemical Society

pression for the thickness of the spreading film. One
concludes that a monomolecular film of oxide is likely
to spread upon the surface of the crystallites driven
by the strong interactions between the oxide species
(TiO_x) and the metal. Various implications of a
physical nature, such as the occurrence of raft-like
crystallites (pillbox morphology), and of a chemical
nature, such as the possibility of an optimum coverage
of the metal surface by TiO_x to obtain maximum activity,
are pointed out.

Supported metal catalysts contain small metal crystallites dispersed
over the internal surface area of refractory metal oxides. The
traditional view was that the support constitutes an inert carrier
whose role is merely to ensure a high dispersion of the metal.
However, while the commonly used refractory oxide supports, silica
and alumina, increase the metal dispersion, they are not inert,
especially toward the non-noble metals and less conspicuously also
toward the noble metals. The physical and chemical interactions
between the active metal, the oxide support and the environment
affect the surface properties of the catalyst and consequently influence the shape of crystallites and the particle size distribution.
Two sets of experimental observations involving surface phenomena are
of interest in the present context: (1) The average size of the
crystallites increases in time, thereby decreasing the surface area
of metal exposed to the chemical atmosphere. This sintering process,
reviewed recently in Reference 1, occurs in order to decrease the
free energy of the system and is therefore affected by the interfacial free energies involved and hence by the interactions (both
physical and chemical) between the crystallites, substrate and
atmosphere. (2) Tauster and Fung (2,3) observed that when Group
VIII metals are supported on TiO_2 and the catalysts reduced at 500°C,
their normal ability to chemisorb H_2 and CO at room temperature is
almost completely suppressed. This suppression is reported to be
absent when the catalysts are prereduced at the lower temperature of
200°C. This Tauster effect was explained by the formation of a
metal-metal bonding between the reduced cations of titanium and any
of the Group VIII metals. An alternate explanation was associated
with the blockage of the metallic surface by TiO_x species resulting
from the reduction of the TiO_2 substrate (4-6).
    The scope of the present paper is to emphasize that the inter-
actions between support, metal and atmosphere are responsible for
both the physical (size distribution, shape of the crystallites,
wettability of the substrate by the crystallites and vice versa), the
chemical and the catalytic (suppression of chemisorption, increased
activity for methanation, etc.) manifestations of the supported metal
catalysts. In the next section of the paper, a few experimental
results concerning the behaviour of iron crystallites on alumina are
presented to illustrate the role of the strong chemical interactions
between the substrate and the compounds of the metal formed in the
chemical atmosphere. Surface energetic considerations, similar to
those already employed by the author (7,8), are then used to explain
some of the observed phenomena. Subsequently, the Tauster effect is
explained as a result of the migration, driven by strong interactions,

of TiO_x species over the surface of the metal and thermodynamic
arguments are adduced to support the premise that the layer of TiO_x
on the metal is monomolecular in thickness. Finally, the insight
gained from the above considerations is used to explain the occurrence of extended planar shapes of the crystallites on the substrate
(pillbox morphology) and to point out some catalytic implications of
the spreading of TiO_x over the surface of the metal.

Experimental

Numerous experiments have been carried out in this laboratory using
particularly alumina as substrate (but also TiO_2) and Pt, Pd, Ni, Co,
Fe and some of their alloys as metals (8–12). Only a few results
obtained with Fe/Al_2O_3 are reported here; more details are included
in Reference 8. The experiments are based on transmission electron
microscopy, which provides information on the physical behaviour of
the crystallites, and on electron diffraction, which provides information on the chemical compounds formed as a result of the interactions between atmosphere, substrate and metal.

Changes occurring in model Fe/Al_2O_3 catalysts were recorded
upon their heating in hydrogen or oxygen atmospheres at temperatures
ranging from 300 to 900°C. Two kinds of hydrogen have been employed
for reduction. In some experiments, ultrahigh pure hydrogen, which
contains <1ppmO_2 and <3ppmH_2O, was used as supplied; in others, this
hydrogen was further purified with the help of a Deoxo unit (Engel-
hard Industries), followed by a silica gel column and by molecular
sieve beds immersed in liquid nitrogen. Samples of 6, 7.5, 10 and
12.5Å initial metal thicknesses deposited by evaporation were
investigated.

Considerable chemical interaction between the traces of oxygen
and moisture in the as supplied hydrogen, iron crystallites and
alumina substrate was detected. In addition to α-Fe in traces, or
Fe_3O_4 (γ-Fe_2O_3 has the same lattice constant), which were detected
when heating in as supplied H_2 or O_2, respectively, solid solutions
leading to $FeAl_2O_4$ and/or $Al_2Fe_2O_6$ were detected in all the cases.
In as supplied H_2, at the low loading of 6 or 7.5Å, very small
amounts of zero valent iron forms even at 700°C, whereas almost
complete reduction to metalic iron is achieved even at 500°C when
the loading is higher, i.e., about 12.5Å. In contrast, when the
as-supplied hydrogen is additionally purified, zero valent iron is
detected even at 400°C and at the low loading of 7.5Å.

Particle migration, coalescence of particles, disappearance of
a large number of small as well as large particles, etc. were
detected. Unusual shapes such as the torus (Figure 1b) and the core-
and-ring structure (Figure 1a) form alternately on heating a sample
in as-supplied hydrogen at 400 or 500°C. The former shape is found
to be associated with a more oxidized state and the latter with a
relatively less oxidized (or relatively reduced) state (8). On
oxidation, at temperatures between 400 and 600°C, considerable
extension of the crystallites and the formation of torus shaped
particles which enclose large cavities were observed. In some cases,
the cavity tends to be filled in, and in other cases there remains an
immobile, randomly located small crystallite in the cavity. On
heating in O_2 at 700°C, following heating at the same temperature in

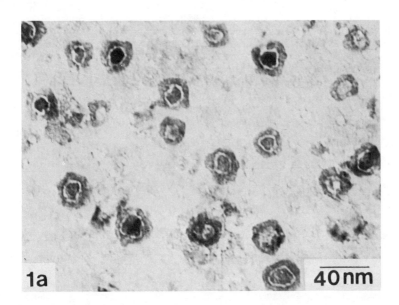

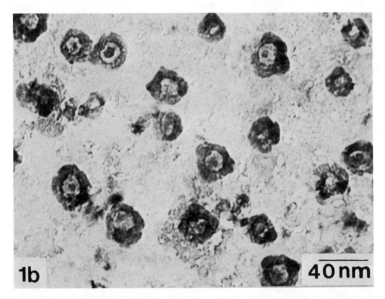

Figure 1. Sequence of changes in the same specimen following heating in as-supplied hydrogen at 400°C for a) 12h and b) additional 6h. The torus (b) corresponds to a more oxidized state, while the core and ring structure (a) corresponds to a relatively less oxidized state.

additionally purified hydrogen, considerably extended crystallites
without any cavity were observed. The heating in O_2 at $700°C$, subsequent to heating in either as-supplied or further purified hydrogen
appears to give rise to a multilayer film coexisting with the three
dimensional crystallites, as inferred from the partial blocking of
the substrate grain boundaries in the micrographs. However, at the
higher temperatures of $800-900°C$ such thick films can be directly
observed (Figure 2). Splitting is observed, both at low and high
loadings, on heating in hydrogen subsequent to heating in oxygen
(Figure 3).

Discussion

A large number of phenomena such as sintering, wetting, shape changes,
splitting, etc., occur during heating of iron-on-alumina catalysts in
H_2 or O_2 environments. Some of these phenomena are discussed below:

Sintering: The traditional mechanisms proposed for sintering are
(1): migration of the crystallites and their coalescence (13),
Ostwald ripening (14-17) and direct ripening (18). Additional
mechanisms of sintering, inferred from our experiments, are a result
of wetting and spreading (19). Small crystallites and even some of
the larger ones extend considerably over the substrate, particularly
during heating in oxygen, into patches of thin (multilayer) films
undetectable by electron microscopy. These patches may contact
either one another or films surrounding the particles. During subsequent heating in hydrogen, the extended patches either coalesce with
the existing crystallites or form, by further contraction, new crystallites located in different places than the initial ones. Of
course, during the extension of the crystallites, undetectable films
around neighboring particles can come into contact to produce necks
that favor the subsequent coalescence of the crystallites. Another
possible scenario is the following: A number of particles extend
during heating in oxygen and form a more or less contiguous (multilayer) film. During heating in H_2, this film being thin, becomes
unstable, for reasons discussed in Reference 19, to thermal or
mechanical perturbations of the free interface and ruptures into a
number of particles which are located in different positions than the
initial particles.

Wetting: The ability of a crystallite to wet a substrate is determined by the following interfacial free energies: substrate-gas
(σ_{sg}), crystallite-gas (σ_{cg}) and crystallite-substrate (σ_{cs}). The
equilibrium contact angle θ of the crystallite on the substrate is
provided by Young's equation:

$$\sigma_{sg} - \sigma_{cs} = \sigma_{cg}\cos\theta \qquad (1)$$

When $\sigma_{sg} - \sigma_{cs} > \sigma_{cg}$, no wetting angle can exist and the crystallite
completely spreads over the substrate. The crystallite has, however,
a finite wetting angle ($0° < \theta < 180°$) when $\sigma_{sg} - \sigma_{cs} < \sigma_{cg}$.
    Under vacuum and in an inert atmosphere, the metals used as
catalysts have high σ_{cg} values as well as high interfacial free
energies σ_{cs} with the commonly used substrates (silica, alumina).

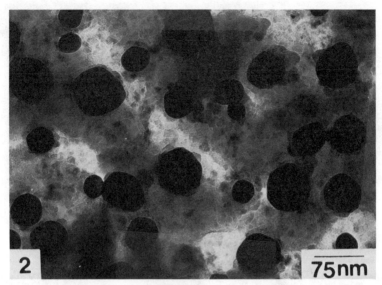

Figure 2. Micrograph showing the presence of a thick film on
the substrate and around the crystallites following heating in
O_2 at 900°C.

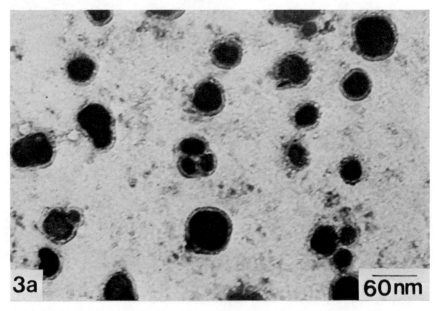

Figure 3 (a). Micrographs depicting the splitting sequence:
a) 2h, hydrogen, 500 C.

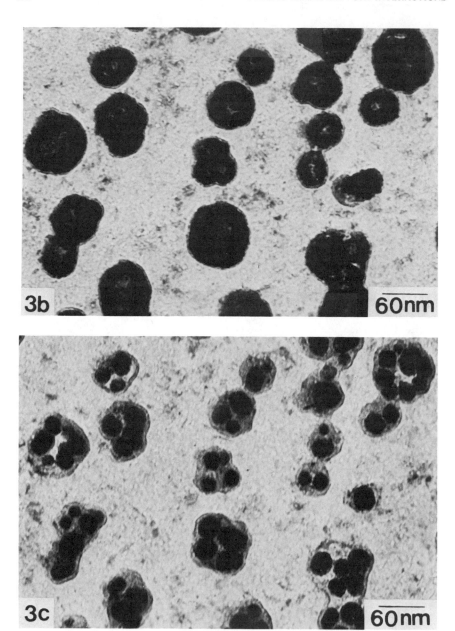

Figure 3 (b,c). Micrographs depicting the splitting sequence:
b) 1h, oxygen, 500 C; c) 1h, hydrogen, 500 C.

This leads to values of θ greater than 90° and the metal does not wet the substrate well. In a hydrogen atmosphere, the chemisorption of the gas on the surface of the metal decreases the value of σ_{cg}. However, in an oxidizing atmosphere, because of the oxidation of the metal, σ_{cg} and σ_{cs} become much smaller than in the above mentioned cases and the contact angle θ can decrease substantially. The interaction energy between the two solid phases in contact plays a major role in this decrease and also in many of the phenomena described in the Introduction and the Experimental parts. The interaction energy and the interfacial free energy between the crystallite and substrate are related via the expression $(\underline{7},\underline{8})$:

$$\sigma_{cs} = \sigma_c + \sigma_s - (U_{int} - U_{str}) \equiv \sigma_c + \sigma_s - U_{cs} \qquad (2)$$

where σ_c and σ_s are the surface free energies of the crystallite and substrate respectively, U_{int} is the interaction energy per unit area of the crystallite-substrate interface between the atoms or molecules of the crystallite and those of the substrate and U_{str} is the strain energy per unit interfacial area due to the mismatch of the two lattices. σ_{cs} between an oxidized crystallite and an oxide substrate is smaller than that between the metal crystallite and the oxide substrate because the interactions between the two oxides are stronger than those between the metal and oxide (like interacts stronger with like than with unlike) and σ_c of the oxidized crystallite is also smaller than the σ_c of the metal. When, in addition, a chemical interaction takes place at the crystallite-substrate interface with the formation of a chemical compound, U_{cs} can become very large, thus decreasing the value of σ_{cs} enormously. If this chemical bonding is very strong, then σ_{cs} could decrease to zero or even below zero under non-equilibrium (dynamic) conditions (at equilibrium, an interface is stable only if $\sigma_{cs} > 0$). This can happen for iron on alumina in an oxidizing atmosphere. The large decrease in the dynamic value of σ_{cs} and also the simultaneous decrease in the value of σ_c (in an oxidizing atmosphere the metal forms an oxide whose σ_c is smaller than the σ_c of the metal) can tremendously increase the driving force for spreading, namely, $\sigma_{sg} - \sigma_{cs} - \sigma_{cg}\cos\theta'$. Here θ' is the instantaneous dynamic wetting angle. The reaction occurs initially at the interface between the substrate and crystallite and propagates into the substrate and/or into the crystallite, most likely into the former. It continues until the rate of dissolution of the oxide molecules becomes limited by a large diffusional resistance. The value of U_{cs} is, however, very large just when the chemical compound, iron aluminate in the present case, forms at the interface. As soon as some of the oxide has already dissolved into the substrate, the subsequent surface interaction energy between aluminate and oxide becomes smaller, since, unlike alumina and oxide, aluminate and oxide do not lead to a chemical compound at the interface. As a result of this, the plot of U_{cs} against time passes through a maximum and σ_{cs} against time passes through a minimum (which could be negative). Consequently, σ_{cs} can have a large decrease initially (becoming possibly even negative) and then increases to an asymptotic value. The large driving force for spreading corresponding to the maximum of U_{cs} leads to a rapid extension of the crystallites. The considerably extended crystallite reaches a new equilibrium angle with the

substrate, at which instant the driving force for wetting becomes
zero. The torus probably forms because the driving force for wetting
is sufficiently large to allow the propagation and coalescence of
cracks existing on the surface of the crystallites. During heating
in as-supplied hydrogen, the crystallites extend on the substrate
because of the oxidation of iron by traces of oxygen and the reaction
between alumina and the iron oxide thus formed. Subsequent to the
oxidation of the metal and its reaction with the substrate, the
crystallites are partly reduced by the excess hydrogen and hence
contract. These extension and contraction cycles become sluggish
eventually (because of fatigue in the system). The extensions during
heating in oxygen and contractions during heating in hydrogen are
caused by the changes in the interfacial free energies in these
atmospheres.

The Tauster Effect

Arguments are provided below to emphasize that the Tauster effect is
also a result of surface phenomena induced by strong interactions.
Two explanations, which are considered opposite to one another, have
been suggested for the Tauster effect. In one of them, the suppression of chemisorption is assumed to be a result of the formation of a
strong metal-metal bonding because of electron transfer between the
substrate and metal. In the other, TiO_x species formed through the
catalytic reduction of TiO_2 are considered to migrate over the surface of the metal. In reality, the second explanation does not exclude the first, since the spreading of TiO_x over the surface of the
metal is triggered by the strong interactions between metal and TiO_x.
The available experimental facts are compatible with the spreading of
a layer of TiO_x species over the surface of the metal. For this
reason, the problem of the spreading of an oxide over a metal will be
examined in some detail. First, in order to use traditional surface
thermodynamics, the film will be assumed to be thick. While not
entirely realistic, this approach provides, nevertheless, useful insight into the Tauster effect. A more realistic treatment that follows considers that the film is thin and provides the possibility of
calculating the thickness of the spreading film from the condition of
the minimum of the free energy with respect to the film thickness.
As noted later in the paper, the main conclusion of the treatment
based on thick films is compatible with the latter more realistic
treatment.

Spreading of a Thick Layer

Metals generally have a much larger surface free energy than the
oxides; therefore, metals do not wet the oxide substrates. However,
for the same reason, an oxide can wet the surface of the metal better.
A thick film of oxide will spread upon the surface of the metal if
the surface free energies σ_{ij} satisfy the following inequality:

$$\sigma_{oxide-metal} + \sigma_{oxide-gas} < \sigma_{metal-gas} \qquad (3)$$

Expressing $\sigma_{oxide-metal}$ in terms of the other two surface free
energies by means of Equation 2, inequality (3) is replaced by:

$$U_{cs} > 2\sigma_{oxide-gas} \tag{4}$$

Inequality (4), which conveys that the spreading of the oxide over the surface of the metal will occur if the interactions per unit interfacial area exceed twice the interfacial free energy of oxide-gas, is both simple and illuminating. Indeed, inequality (4) is satisfied if $\sigma_{oxide-gas}$ is sufficiently small and/or U_{cs} is sufficiently large. From the data available in the literature (20) on the surface free energies of oxides, those pertinent to the oxides used as catalytic substrates are summarized in Table I.

TABLE I

	OXIDE	σ ergs/cm^2	T°K	σ ergs/cm^2 at T=298°K
NON-SMSI	ZrO$_2$	1130 770	<1423 1423-2573	1243
	MgO	1000	298	1000
	Al$_2$O$_3$	680	2323	883
	SiO$_2$	605	298	605
SMSI	TiO$_2$	380	2100	560
	Ta$_2$O$_5$	280	2100	460
	V$_2$O$_5$	90	963	156

The values in the last column have been calculated assuming $\frac{d\sigma}{dT} = -0.1$ ergs/cm^2°C

It is interesting to note that the oxides which do not manifest the SMSI, such as ZrO$_2$, MgO, Al$_2$O$_3$ and SiO$_2$, have large values; those which do manifest SMSI, such as TiO$_2$, Ta$_2$O$_5$, V$_2$O$_5$, have low values. Of course, it is easier to satisfy inequality (4) if $\sigma_{oxide-gas}$ is small, but a more complete comparison should also include the interactions between the oxide and metal. If the interactions between oxide and metal are strong, for instance of the order of 60 Kcal/mole, then, if one assumes 10^{15} species of TiO$_x$ per cm^2, one obtains $U_{cs} \simeq 3300$ dyne/cm. Comparing this figure with Table I, it is clear that inequality (4) can indeed be satisfied when strong interactions between metal and substrate occur.

   Even though thermodynamics ensures that spreading will occur if inequality (4) is satisfied, the rate of the process may be too slow for this phenomenon to be observed. In order to enhance the rate of the process, either vacancies have to be created and/or bonds broken

to increase the mobility of the substrate species. This can be
achieved by depleting the oxide of some oxygen, by reduction for
example as has been generally carried out. However, depletion could
also be achieved by heating in vacuum or even in an inert atmosphere,
although at higher temperatures of course. In addition, one may note
that: (1) the strong interactions between metal and TiO_x can increase tremendously the value of U_{cs}, thus ensuring that inequality
(4) is satisfied, and (2) while it is likely that TiO_2 has a lower
σ_{sg} value than TiO_x, the stronger interactions between the more
mobile and reactive species TiO_x and metal probably overcompensate
for its increased surface free energy.

Spreading of a Thin Layer

The real problem, however, is to predict the thickness of the film.
In the problem on hand, a number of crystallites are distributed over
the surface of the substrate and some of the molecules of the substrate spread upon the crystallites because the strong interactions
can decrease the free energy. The thickness of the spread layer
should be such as to minimize the free energy of the system. The
main difficulty is to derive an expression for the free energy of
formation of a thin film. In a thick film, the majority of the
molecules have a range of interaction smaller than the thickness of
the film, whereas in a thin film, the molecules have a range of
interaction larger than the film thickness. For this reason, the
free energy of formation of a thin film depends on its thickness,
while that of a thick film does not. For the sake of simplicity and
for illustrative purposes, the Lennard–Jones expression is used for
the interaction potential between the oxide molecules themselves and
between the oxide molecules and those of the metal. The following
expression is then obtained for the free energy of formation of a
thin film of thickness h on a substrate (21):

$$\sigma = \sigma_\infty + \frac{\alpha}{h^2} - \frac{\beta}{h^8} \qquad (5)$$

where

$$\sigma_\infty = \sigma_{oxide-metal} + \sigma_{oxide-gas} - \sigma_{metal-gas} \qquad (6)$$

and α and β, which are both negative quantities, depend upon the coefficients in the Lennard–Jones potentials (oxide–oxide and oxide–
metal). The second term in the right hand side of Equation 5 is due
to the van der Waals interactions, while the third term is due to the
Born repulsion. Since a continuum approach is used in its derivation,
Equation 5 holds only for film thicknesses that are sufficiently
large compared to molecular sizes; nonetheless, for qualitative purposes it will be extrapolated also to small thicknesses. While the
above equations have been derived in some detail in Reference 21, it
is instructive for the present purpose to clearly define the free
energy of formation σ. This quantity represents the difference between the free energy due to the interactions between the molecules
of the film and those of the metal and the free energy due to the
interactions between the molecules of the film and the substrate

before their migration. This means that migration from the substrate to the surface of the metal will occur if $\sigma<0$ and that the
thickness h_m of the spread film is such as to make σ minimum. Using
Equation 5, one obtains that

$$h_m = \frac{(4\beta)^{1/6}}{\alpha^{1/6}} \tag{7}$$

and that spreading will occur if

$$\sigma(h_m) = \sigma_\infty + \frac{3}{4^{4/3}} \frac{\alpha^{4/3}}{\beta^{1/3}} < 0 \tag{8}$$

Because α and β are negative quantities, the second term in the right
hand side is also a negative quantity and this has interesting implications.   It is clear from Equation 8 that if σ_∞ is negative, $\sigma(h_m)$
will also be negative.   However, $\sigma(h_m)$ can have a negative value even
for positive values of σ_∞, if the second term is sufficiently negative and/or σ_∞ is not too large.   One may also note that the condition $\sigma_\infty<0$ coincides with the criterion for spreading of a thick film.
Of course, the thickness of the spreading film has to result from the
condition of minimum of the free energy.   However, the condition $\sigma_\infty<0$
is more stringent than the condition provided by Equation 8 and
therefore, any result derived from the former, such as inequality (4)
will also be more stringent.   This means that if $U_{cs}>2\sigma_{oxide-gas}$
spreading will surely occur; it can, however, also occur even when
U_{cs} is somewhat smaller than $2\sigma_{oxide-gas}$.
    While Equation 5 is approximate and involves an interaction
potential which is not valid for strong interactions, it can still
illuminate the consequences of strong interactions for the following
reason.   The thickness h given by Equation 7 is determined both by
the van der Waals interactions and Born repulsion and obviously has
to be smaller than the range of Born repulsion which is smaller than
say a few Angstrom.   In the case of strong interactions, the two dissimilar species will be in very close contact, which means that the
repulsive interactions will have a shorter range.   This line of
reasoning in such cases suggests that a monolayer (or submonolayer)
of TiO_x is likely to spread upon the surface of the metal.
    It is important to note that the cases of metal crystallites on
a substrate and of a substrate of arbitrary thickness deposited upon
a metal foil are not equivalent from a thermodynamic point of view
because the constraints to which each of these systems are subjected
are different.   In the first case, a monolayer of TiO_x will cover the
metal, the amount being determined by the equilibrium with the surface of the substrate.   For the second, the entire deposit of TiO_2
must be located on the surface.   Since the coverage by a monolayer
leads to the smallest free energy, the excess of TiO_x should form in
the latter case a tridimensional structure with the least possible
surface area over the smallest possible part of the substrate
surface, thus minimizing the free energy.   There are, however,
kinetic difficulties to achieve such a structure.   For this reason,
if $\sigma_\infty<0$, it is likely that a metastable state of extended patches

will form which although less stable thermodynamically, can be more easily achieved kinetically. These patches should have, for reasons discussed later in the paper, an abrupt variation of angle near their leading edges. If $\sigma_\infty>0$ and $\sigma<0$, then instead of a single crystallite, several crystallites will coexist with the monolayer, again because such a state is more easily achievable kinetically.

The above considerations are based on expression (5). If more complete expressions are used for the interaction potential, the dependence of σ on h will be more complicated, resulting in additional manifestations of the spreading phenomena. This will be taken up in a later paper.

Mechanism of Migration of the TiO_2 Species

The migration of the TiO_x species (probably as a monolayer) upon the surface of the metal occurs because this decreases the free energy of the system; in other words, TiO_x constitutes a surface active agent for the surface of the metal. The gradients of the surface concentration and of the surface free energy constitute the driving forces of the process. Molecules of TiO_x are adsorbed at the leading edge of the crystallite and the adsorbed molecules move from the leading edge, where their concentration is larger and the surface free energy is smaller, in the direction of increasing surface free energies. Indeed, if there is a variation of the surface free energy along an interface, then the interface has the tendency to relapse into a state with a minimum free energy through the expansion of the regions with a low surface free energy and contraction of those with a high surface free energy. The gradient of the surface free energy along the surface gives rise to a shear stress and the molecules of TiO_x migrate along the surface under the action of this stress and of the concentration gradient. At thermodynamic equilibrium, the fraction of the metal surface that is occupied can be calculated from the equality of the chemical potentials of TiO_x on the substrate, μ_s, and on the surface of the metal, μ. Neglecting the interactions between the TiO_x molecules on the surface of the metal as well as the effect of the curvature in comparison with the strong interactions between TiO_x and metal, one can use for the chemical potential μ the expression valid for the Langmuir isotherm:

$$\mu = kT\ln \frac{\theta}{(1-\theta)q}$$

where k is the Boltzmann constant, T is the absolute temperature, q is the partition function for a single adsorbed molecule and θ is the fraction of sites occupied. Consequently

$$\theta = \frac{qe^{\mu_s/kT}}{1+qe^{\mu_s/kT}} \tag{9}$$

Additional assumptions regarding μ_s and q are however, needed to express θ as a function of temperature. For strong interactions q is very large and one can therefore expect θ to be close to unity.

Strong Interactions and the Shape of the Crystallites

When Pt supported on TiO2 is heated in H2, the Pt crystallites spread
to form thin planar structures, hexagonal in shape (22). Similar results have been also obtained in other cases (23-25). This appears
to be a result of the reduction of TiO2 to TiO$_x$ in the presence of
Pt. This reduction increases the surface free energy σ_{sg}. However,
the surface free energy σ_{cg} decreases (because of the migration of
TiO$_x$ over the surface of the metal and the strong interactions between the two, and (as a result of the same interactions) the interfacial free energy σ_{cs} also decreases. Therefore, $\sigma_{sg} - \sigma_{cs}$ can
become greater than σ_{cg} and hence the crystallite has the tendency to
spread over the substrate. This cannot, however, explain the formation of the planar structure with an abrupt variation of the angle
near the leading edge of the crystallite. The possibility of a
planar structure which exhibits a rapid variation of angle near the
leading edge was anticipated theoretically by Ruckenstein and Lee in
1975, on the basis of the following considerations (26). The thermodynamic approach on which Young's equation is based involves the existence of a macroscopic wetting angle which is defined on a length
scale which is large compared with the atomic dimensions. One can,
however, demonstrate that within a short distance of a few nanometers
from the leading edge, the internal angle between the horizontal and
the line connecting the centers of two successive molecules at the
solid-gas interface varies rapidly from some value θ_0 at the leading
edge to the smaller thermodynamic value θ at some distance from the
leading edge. The calculations have been carried out by using for
the interaction potential, ϕ_{ij}, between two atoms of species i and j
whose centers are at a distance r apart, the (simple) expression:

$$\phi_{ij} = -\frac{\beta_{ij}}{r^6} \text{ for } r > a_{ij} \text{ and } \phi_{ij} = \infty \text{ for } r < a_{ij} \tag{10}$$

where a_{ij} is the minimum distance of approach of the atoms of species
i and j (the hard core radius) and β_{ij} characterizes the strength of
the attractive interactions between species i and j. The rapid variation of the angle is caused in this case by the hard core repulsion
(exclusion) between the atoms of the crystallite and those of the
substrate. At the leading edge, where the atoms of the crystallite
and those of the substrate are in close proximity, the effect of this
exclusion is greatest and consequently the internal angle is the
largest. The macroscopic, thermodynamic angle θ is given by Young's
equation:

$$\sigma_{sg} - \sigma_{cs} = \sigma_{cg}\cos\theta \tag{11}$$

and is achieved asymptotically at some distance from the leading edge.
Since the angle θ_0 at the leading edge is larger, it will be given by
an expression of the form:

$$\chi(\sigma_{sg} - \sigma_{sc}) = \sigma_{cg}\cos\theta_0 \tag{12}$$

where χ is a factor smaller than unity, whose form is given in
Reference 26.

The leading edge angle θ_0 is the angle between the line connecting the centers of two successive atoms located at the leading edge of the crystallite-gas interface and the horizontal. Values of $\cos\theta_0 > 1$ obtained from Equation 12 are incompatible (because they cannot exist) with the presence of a second layer of molecules on the first layer. In this case, the atoms will spread as individual atoms over the substrate. However, if $\cos\theta > 1$ and $\cos\theta_0 < 1$ total spreading can occur at some distance from the leading edge while the atoms near the leading edge still have a finite wetting angle with the substrate. In this case the spreading will generate an extended thin planar crystallite whose profile changes rapidly in a region of a few nanometers near the leading edge. As already noted, strong interactions are needed to achieve $\cos\theta > 1$. If these interactions are, however, too strong, then $\cos\theta_0$ can become greater than unity and the atoms will spread as individual atoms over the surface of the substrate.

Catalytic Implications

In a recent kinetic theory of the selectivity of catalytic processes (27,28) the concept of a spectrum of landing areas was introduced and employed. According to that theory, the molecules can land in a variety of ways upon the surface of the catalyst (even upon one whose surface is energetically homogeneous). Various kinds of landings involve various numbers and arrangements of sites and lead to various products. While a particular site can be a part of several kinds of landing areas, the likelihood of finding a larger number of more extended landing areas for particular products is smaller on a smaller crystallite than on a larger one. This explains the effect of the size on the selectivity of the catalytic process. The interactions between metal, substrate and atmosphere affect the shape, composition and size distribution of the crystallites. For reasons just noted, they will also affect the activity and selectivity of the catalyst. In addition, because of the migration of the substrate species on the surface of the metal during prereduction of the Group VIII metals on TiO2, the spectrum of landing areas will be changed tremendously. The structure insensitive reactions will be only attenuated by the presence of the layer of TiO_x on the metal. In this case, the interactions between the metal and chemisorbed molecules will still take place, but the rate of electron transfer between them is decreased because the electrons have to tunnel through the TiO_x layer. Indeed, for Rh/TiO2, the specific activity for benzene hydrogenation or for the dehydrogenation of cyclohexane is decreased (by a factor of less than two) as the reduction temperature is increased to 500°C (29,30). The change in activity for Ir or Pt is much larger (4,29). The structure sensitive reactions, such as hydrogenolysis, need landing areas on the surface of the metal whose likelihood to exist on a metal surface largely occupied by TiO_x is very small. A major decrease in their activity is therefore expected to occur and this, indeed, occurs (4,30). The reaction between CO and H2 (31-36) is of particular interest because the activity for methanation of Ni/TiO2 is increased by the presence of TiO_x on the metal. This could be a result of the following two opposite effects: On one hand, a large number of Ni sites are covered by TiO_x and therefore there are fewer

sites over which chemisorption can occur. On the other hand, the
presence of TiO_x molecules nearby a metal atom increases the rate of
reaction, either because (1) it decreases the strength of the chemisorption bond with the metal thus increasing the rate of desorption
of the reaction product, (2) TiO_x and the metal form a bifunctional
catalyst which accelerates the reaction, (3) the presence of TiO_x
accelerates the reaction for steric reasons providing a useful tem-
plate for the final product, and (4) combinations of the above.
Irrespective of the real mechanism, the migration of TiO_x decreases
the number of chemisorption sites but increases greatly the reaction
rate on the remaining sites when their number decreases. This means
that by selecting an appropriate prereduction temperature one can
find an optimum coverage which will provide the maximum activity of
the catalyst. The coverage of the metal surface by TiO_x favors the
formation of compounds that need landing areas with fewer metalic
sites. For instance, CO chemisorbs on one or two sites. Because of
the surface exclusion provided by the existance of TiO_x species, the
chemisorption of CO on one site, which is proportional to $1-\theta$ (θ being
the fraction of the surface occupied by TiO_x), is more likely to
occur than that on two sites, which is proportional to $(1-\theta)^2$. This
means that the compounds based on one site chemisorption of CO are
also more likely to form than those based on two sites chemisorption.

Conclusions

Surface thermodynamic considerations can be helpful in an understanding of the complex phenomena which occur in supported metal catalysts.
Indeed, the physical and chemical interactions between metal, substrate and atmosphere lead to wetting and spreading phenomena (of the
active catalyst over the substrate and of the substrate over the
metal) which are relevant for the physical (sintering, redispersion)
as well as chemical (suppression of chemisorption, modification of
selectivity, enhanced activity) manifestations of supported metal
catalysts.

Acknowledgments

The experiments have been carried out by I. Sushumna to whom I am
also indebted for a critical reading of the manuscript.

Literature Cited

1. Ruckenstein, E,;Dadyburjor, D.B. Reviews in Chemical Engineering
   1983, 1, 251.
2. Tauster, S.J.; Fung, S.C.; Garten, R.L. J.A. Chem. Soc. 1978,
   100, 170.
3. Tauster, S.J.; Fung, S.C. J. Catal. 1978, 55, 29.
4. Meriaudeau, D.; Dutel, J.F.; Dufaux, M.; Naccache, C. In
   "Stud. Surf. Sci. Catal." 1982, 11, 95.
5. Santos, J.; Phillips, J.; Dumesic, J.A. J. Catal., 1983, 84, 147.
6. Raupp, G.B.; Dumesic, J.A. J. Phys. Chem., 1984, 88, 660.
7. Ruckenstein, E.; Pulvermacher, B. J. Catal., 1973, 29, 244.
8. Sushumna, I.; Ruckenstein, E. J. Catal., 1985, 94, 239.
9. Chen, J.J.; Ruckenstein, E. J. Catal., 1981, 69, 254.
10. Chen, J.J.; Ruckenstein, E. J. Phys. Chem., 1981, 85, 1696.

11. Sushumna, I.; Ruckenstein, E. J. Catal., 1984, 90, 241.
12. Sushumna, I. Ph.D. Dissertation, 1985.
13. Ruckenstein, E.; Pulvermacher, B. A.I. Ch. E. Journal, 1973, 19, 356.
14. Chakraverty, B.K. J. Phys. Chem. Solids 1967, 28, 2401.
15. Wynblatt, P.; Gjostein, N.A. Prog. Solid State Chem. 1975, 9, 21.
16. Flynn, P.C.; Wanke, S.E. J. Catal. 1974, 34, 390.
17. Ruckenstein, E.; Dadyburjor, D.B. J. Catal. 1977, 33, 233.
18. Ruckenstein, E.; Dadyburjor, D.B. Thin Solid Films 1978, 55, 89.
19. Ruckenstein, E.; Sushumna, I. J. Catal. (in press).
20. Overbury, S.H.; Bertrand, P.A.; Somorjai, G.A. Chem. Rev. 1975, 75, 547.
21. Ruckenstein, E. In "Growth and Properties of Metal Clusters"; J. Bourdon, Ed.; Elsevier: Amsterdam, 1980, p. 37.
22. Baker, R.T.K.; Prestridge, E.B.; Garten, R.L. J. Catal. 1979, 56, 390; 1979, 59, 293.
23. Tatarchuk, B.J.; Dumesic, J.A. J. Catal. 1981, 70, 308, 323 and 335.
24. Tatarchuk, B.J.; Chludzinski, J.J.; Sherwood, R.D.; Dumesic,J.A.; Baker, R.T.K. J. Catal. 1981, 70, 933.
25. Huizinga, T.; Prins, R. J. Phys. Chem. 1981, 85, 2156.
26. Ruckenstein, E.; Lee, P.S. Surf. Sci.1975, 52, 298; J. Colloid Interface Sci. 1982, 86, 573.
27. Ruckenstein, E.; Dadyburjor, D.B. Chem. Eng. Commun. 1982, 14, 59.
28. Dadyburjor, D.B.; Ruckenstein, E. J. Phys. Chem. 1981, 85, 3396.
29. Ellestad, O.H.; Naccache, C. In "Perspectives in Catalysis"; Proceedings 12th Swedish Symposium on Catalysis, Lund, R. Larsson Ed.; CWK Gleerys, Lund.
30. Haller, G.L.; Resasco, D.E.; Ronco, A.J. Faraday Discussion 1982, 72, 109.
31. Ryndin, Yu A.; Hicks, R.F.; Bell, A.T.; Yermakov, Yu I. J. Catal. 1981, 70, 287.
32. Wang, S.Y.; Moon, S.H.; Vannice, M.A. J. Catal. 1981, 71, 167.
33. Solymosi, F.; Tombacz, I.; Kocsis, M. J. Catal. 1982, 75, 78.
34. Vannice, M.A.; Garten, R.L. J. Catal. 1979, 56, 236; 1980, 66, 242.
35. Vannice, M.A. J. Catal. 1982, 66, 242.
36. Kao, C.C.; Tsai, S.C.; Chung, Y.W. J. Catal. 1982, 73, 136.

RECEIVED September 12, 1985

Effects of Reduction Temperature on H_2 Adsorption by Pt on Various Supports

Jacek A. Szymura[1] and Sieghard E. Wanke

Department of Chemical Engineering, University of Alberta, Edmonton, Alberta, Canada T6G 2G6

The influence of reduction temperature on subsequent hydrogen chemisorption for Pt supported on magnesias, aluminas and silicalite is examined in this paper. Significant decreases in hydrogen chemisorption capacities with increasing reduction temperatures were observed for most catalysts. X-ray diffraction results showed that sintering of Pt was not the major cause for the suppression in adsorption uptakes. Poisoning of the Pt surface by substances which originate in the support and which migrate onto the Pt surface during reduction at elevated temperatures is the probable cause for the suppression of adsorption observed for Pt/MgO and Pt/silicalite. Adsorption suppression for Pt supported on sulfur-containing MgO was observed after reduction at temperatures as low as 350°C. For Pt/Al$_2$O$_3$, reduction at 700°C was required before significant decreases in hydrogen adsorption were observed.

Hydrogen chemisorption has been used extensively for over two decades for the determination of metal surface areas, i.e. metal dispersions of supported platinum group metal catalysts. In order to convert measured hydrogen adsorption uptakes to metal dispersions, the amount of hydrogen adsorbed per surface metal atom has to be known, i.e. the adsorption stoichiometry has to be known. Considerable amount of research was carried out in the 1960s and 1970s to determine adsorption stoichiometries for various adsorbates on different metals for a range of adsorption pressures and temperatures (1-7). Based on these and other studies, it was concluded that hydrogen adsorption on platinum at temperatures close to room temperature and low pressures corresponds to approximately one hydrogen atom per surface platinum atom.

[1]Current address: Institute of Technology, Technical and Agricultural Academy, 85-326 Bydgoszcz, Poland

However, in the past few years many investigators have reported
hydrogen adsorption uptakes on supported platinum catalysts which
correspond to adsorption stoichiometries of much less than one hydrogen atom per surface platinum atom. This suppression of hydrogen adsorption is usually brought about by reducing the supported catalyst
in hydrogen at elevated temperatures. Tauster and co-workers (8-9)
were the first to report dramatic decreases in hydrogen adsorption
capacity as a result of high temperature reduction for various platinum metals on reducible supports such as TiO_2, V_2O_3 and Nb_2O_5. They
coined the phrase 'Strong Metal–Support Interactions' (SMSI) to
denote this behavior which was attributed to the reducibility of the
supports. Decreased hydrogen adsorption uptakes as a result of high
temperature reductions have also been reported for platinum supported
on alumina, a much more difficult support to reduce (10-11).
Dautzenberg and co-workers also attributed the suppression in hydrogen
adsorption to metal–support interactions, i.e. the formation of
$Pt-Al_2O_x$ complexes. It is now well established that reduction temperature can have a significant effect on subsequent hydrogen adsorption uptakes (12), but no single explanation for this phenomenon has
been universally accepted.

In the present paper, results on the influence of reduction temperature on subsequent hydrogen adsorption are reported for platinum
supported on three difficult to reduce supports (γ-alumina, magnesia
and silicalite). The results indicate that the observed suppression
in hydrogen uptakes, after high temperature reduction, are due to
poisoning of the platinum surface by species originating from the
support. This interpretation is similar to that of Wang et al. (13).

Experimental Methods

Supports and Catalysts. The catalyst supports used in this work are
described in Table I. The surface areas, except for the silicalite,
were measured by the multi-point BET method. The surface area for
the silicalite was obtained from the manufacturer. Silicalite is an
essentially aluminum-free pentasil zeolite (14) manufactured by Union
Carbide. The chlorine contents of the supports were determined by
neutron activation analysis, and sulfur contents were obtained with a
Leco sulfur analyser. Sulfur and chlorine contents were measured
since these elements may influence subsequent hydrogen adsorption on
the supported platinum catalysts (15).

Support MgO-3 was obtained by steam stripping MgO-2 at
atmospheric pressure for 40 h at 500°C, 105 h at 550°C and 110 h at
600°C, followed by dehydration in flowing, dry nitrogen at 600°C for
4 h. This treatment reduced the chlorine content from 0.41 to 0.03
wt% (see Table I). Support Al_2O_3-2 was obtained by steam stripping
Al_2O_3-1 at atmospheric pressure for 24 h at 350°C and 24 h at 400°C.
This treatment reduced the chlorine content from 1.5 to 0.04 wt%.

Supported platinum catalysts, described in Table II, were
prepared by impregnation of MgO and Al_2O_3 supports and by exchange for
the silicalite support. Impregnations were done with aqueous
solutions of hexachloroplatinic acid and with acetone solutions of
platinum acetylacetonate. The impregnation procedures with aqueous
chloroplatinic acid have been described previously (16). The impregnation procedure with $Pt(C_5H_7O_2)_2$ – acetone consisted of wetting the
support with acetone, addition of $Pt(C_5H_7O_2)_2$– acetone solution,

Table I. Description of Supports

Support	Surface Area (m²/g)	Chlorine Content (wt%)	Sulfur Content (wt%)
MgO-1	30	<0.05	0.32
MgO-2	90	0.41	0.02
MgO-3	--	0.03	0.02
MgO-4	16	<0.001	n.d.*
Al₂O₃-1	100	1.5	n.d.
Al₂O₃-2	95	0.04	n.d.
Silicalite	430†	<0.005	n.m.#

* n.d. = not detected; # n.m. = not measured
† area obtained from supplier (single point BET)

Table II. Description of Catalysts

Catalyst	Support	Platinum Precursor	Pt Loading (wt %)
CAT-1	MgO-1	H_2PtCl_6	0.48
CAT-2	MgO-2	H_2PtCl_6	0.59
CAT-3	MgO-2	$Pt(C_5H_7O_2)_2$	0.56
CAT-4	MgO-2	H_2PtCl_6	5.9
CAT-5	MgO-2	$Pt(C_5H_7O_2)_2$	5.6
CAT-6	MgO-3	$Pt(C_5H_7O_2)_2$	0.55
CAT-7	MgO-4	H_2PtCl_6	0.60
CAT-8	MgO-4	$Pt(C_5H_7O_2)_2$	0.53
CAT-9	Al₂O₃-1	$Pt(C_5H_7O_2)_2$	1.06
CAT-10	Al₂O₃-2	H_2PtCl_6	1.27
CAT-11	Al₂O₃-2	$Pt(C_5H_7O_2)_2$	1.07
CAT-12	Silicalite	$Pt(NH_3)_4Cl_2$	0.98

intermittent stirring during next 24 h, evaporation of excess acetone
at 35°C, and drying at 75°C for 24 h. The Pt/silicalite catalyst
(CAT-12, Table II) was prepared by immersing the silicalite in an
aqueous solution of $Pt(NH_3)_4Cl_2$ at 22°C for 60 h, followed by
filtering the solids from the solution, washing the solids with
deionized distilled water (eight times) and drying in air at 300°C
for 3 h.
    All the dried catalysts were reduced in flowing hydrogen at
150°C for 16 h and 250°C for 2 h. The reduced catalysts were stored
in air until use. The platinum contents listed in Table II are on a
dry basis and were determined by neutron activation analysis.

Treatment and Characterization of Catalysts. A dynamic pulse adsorption apparatus (16) was used to carry out in situ pretreatments, reductions, degassing and hydrogen chemisorption measurements.
Portions of the prereduced, air-stored catalysts (1 to 3 g for the
0.5 to 1.0 wt% Pt catalysts and 0.3 to 0.4 g for the 5 to 6 wt % Pt
catalysts) were placed into a quartz U-tube which was then attached
to the adsorption apparatus. Each catalyst sample was subjected to
various sequential oxygen treatments and reduction conditions. After
the desired reduction, usually for 1 h, the catalyst was degassed in
ultra-pure nitrogen before hydrogen adsorption uptakes were determined at 22°C. All the treatments (oxygen, reduction and degassing)
were done in flowing gases at atmospheric pressure.

        After many of the adsorption measurements the sample was removed
from the U-tube and examined by wide-angle X-ray diffraction (XRD).
A Philips X-ray diffractometer using the step-scan mode and Cu $K\alpha$
radiation was used to obtain the XRD patterns in the range of 36 to
44° of 2θ. The step size used was 0.02° of 2θ with a counting time
of 100 s per step. Comparison of XRD results with transmission electron microscopy results showed that at the above scan rates the Pt
111 line at 39.8° was readily detected for samples containing Pt
crystallites ≥2.0 nm even for Pt contents of 0.5 wt%.

Results and Discussion

Platinum Supported on Magnesia. Results of hydrogen adsorption as a
function of reduction temperature for Pt on various MgO supports are
tabulated in Table III. For CAT-1 to CAT-6 increases in reduction
temperature resulted in significant decreases in subsequent room temperature hydrogen adsorption uptakes. The hydrogen adsorption capacities for catalysts reduced at 500°C were restored by oxygen treatment at 550°C followed by low temperature (250 to 300°C) reduction
(cf. Runs 4 and 5, and Runs 7 and 8, Table III). Similar
reversibilities in suppression and restoration of hydrogen adsorption
capacities were observed for CAT-3 to CAT-6. The results for Runs 6
to 9 show that the high-low-high hydrogen adsorption sequence occurs
for many cycles of low temperature reduction – high temperature reduction – O_2 treatment followed by low temperature reduction.

        It is unlikely that this reversibility is due to reversible
changes in Pt dispersion. The XRD results for CAT-4 shown in Figure
1 clearly show that increasing the reduction temperature from 300 to
500°C does not result in appreciable changes in Pt dispersion. The
XRD patterns after 300°C reduction (Runs 12 and 14) are very similar
to the corresponding patterns after 500°C reduction (Runs 13 and 15).
The similarity of the patterns is shown in Figure 1B. The subtracted
patterns in Figure 1B show slight maxima at 40°, and minima and
maxima at about 43°. The minima and maxima at 43° are due to very
small mismatches in the angles and the sharpness and intensity of the
MgO 200 line, i.e. if the patterns for Runs 13 and 15 are shifted by
0.02° before subtraction then the minima and maxima are reversed.
The small maxima at about 40° may be due to slight increases in Pt
crystalite size as a result of reduction at 500°C. These small
decreases in Pt dispersion, however, cannot account for the over
five-fold decrease in hydrogen adsorption after the 500°C reductions
(cf. H_2 uptakes for Runs 12 and 13 and Runs 14 and 15, Table III).
XRD results for all the other Pt/MgO catalysts also showed that in-

creasing reduction temperatures from 300 to 500° did not result in appreciable changes in Pt dispersion.

However, treatment in oxygen at 550°C of Pt/MgO resulted in significant changes in Pt dispersion. This is clearly shown in Figure 1 (cf. XRD intensities of the Pt 111 lines for Runs 12 and 14). Previously reported electron microscopy results have also shown that oxygen treatment at 550°C results in decreases in Pt crystallite sizes (17). The suppression of hydrogen adsorption as a result of 500°C reduction occurred for both low and high Pt dispersions, and hence, is not a strong function of Pt dispersion.

However, not all the Pt/MgO catalyst displayed reduced hydrogen uptakes after high temperature reduction; reduction temperature did not influence hydrogen uptakes for CAT-7 and CAT-8 (see Table III). The only major difference between these two catalysts and the previously discussed six catalysts was the sulfur content; CAT-7 and CAT-8 were essentially sulfur-free (see Table I). The chlorine content and platinum precursor do not appear to influence hydrogen adsorption uptakes. Hence, it is concluded that sulfur in the MgO is responsible for the decrease in hydrogen adsorption with increasing reduction temperature. Similar results were obtained by Wang et al. (13) with Rh supported on MgO.

Table III. Effect of Pretreatment on Hydrogen Adsorption for Pt/MgO

Catalyst	Support	O_2 Treatment*	Run	Reduction Temp.(°C)	H_2 Uptake (H/Pt)
CAT-1	MgO-1	16 h at 550°C	1	250	0.26
			2	350	0.10
			3	400	0.01
			4	500	0.00
		16 h at 550°C	5	250	0.26
CAT-2	MgO-2	16 h at 550°C	6	300	0.27
			7	500	0.03
		see below†	8	300	0.23
			9	500	0.01
CAT-3	MgO-2	16 h at 550°C	10	300	0.26
			11	500	0.00
CAT-4	MgO-2	none	12	300	0.15
			13	500	0.02
		1 h at 550°C	14	300	0.42
			15	500	0.08
CAT-5	MgO-2	16 h at 550°C	16	300	0.57
			17	500	0.25
CAT-6	MgO-3	1 h at 550°C	18	300	0.11
			19	500	0.00
CAT-7	MgO-4	1 h at 550°C	20	300	0.16
			21	500	0.19
CAT-8	MgO-4	1 h at 550°C	22	300	0.10
			23	500	0.10

* Treatments for each catalyst done sequentially on the same sample.
† Four O_2-H_2-N_2 treatment cycles at 550°C done before reduction.

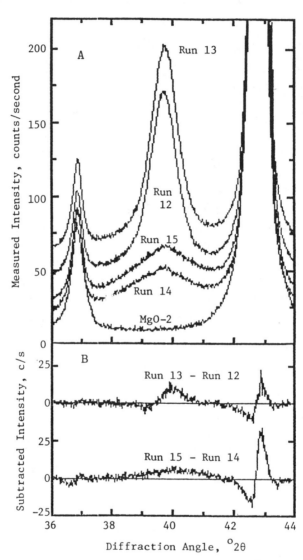

Figure 1.   XRD patterns for CAT-4 (5.9% Pt/MgO-2).
A: Patterns as measured (offset for clarity);
B: Subtracted patterns (see Table III for treatments).

The reversibility of this phenomenon indicates that during high
temperature reduction a reduced sulfur species is formed which
migrates to the Pt surface and poisons the Pt surface. Subsequent
oxygen treatment oxidizes this sulfur species and the oxidized
species migrates back to the support. The oxidized species do not
leave the catalyst since the reduction - oxidation sequence produces
the same effects over several cycles (see Runs 6 to 9, Table III).
If, for example, it is assumed that high temperature reduction of
CAT-2 causes 50% of the Pt surface to be covered by sulfur and that
all this sulfur leaves the catalyst during the high temperature oxidation, then less than two high temperature reduction - oxidation
cycles would be required to remove all the sulfur (0.02 wt%) from the
MgO, but reduced hydrogen uptakes still occurred after five
oxidation-reduction cycles (see Run 9).

Platinum Supported on Alumina.   Reduction of $Pt/\gamma-Al_2O_3$, both Al_2O_3-1
and -2 are $\gamma-Al_2O_3$, at high temperatures also results in marked
decreases in hydrogen adsorption uptakes. However, reductions at
temperatures above 500°C are required to cause significant
suppressions in hydrogen adsorption. Hydrogen adsorption results
which illustrate this behavior are presented in Table IV. XRD results in Figure 2 show that the low H/Pt ratios after 700°C reduction
cannot be attributed to decreases in Pt dispersions.

Table IV.   Effect of Pretreatment on Hydrogen Adsorption for Pt/Al_2O_3

Catalyst	Support	O_2 Treatment*	Run	Reduction Temp.(°C)	H_2 Uptake (H/Pt)
CAT-9	Al_2O_3-1	1 h at 550°C	24	300	1.04
			25	500	0.79
CAT-10	Al_2O_3-2	1 h at 550°C	26	300	1.13
			27	500	0.83
			28	700	0.33
			29	700†	0.19
		0.5 h at 22°C	30	500	0.49
		0.5 h at 100°C	31	500	0.55
		0.5 h at 200°C	32	500	0.55
		0.5 h at 300°C	33	500	0.55
		1.0 h at 700°C	34	500	0.22
CAT-11	Al_2O_3-2	1.0 h at 550°C	35	300	0.32
			36	500	0.31
		16 h at 550°C	37	300	0.29
			38	500	0.26
			39	700	0.03
		24 h in air	40	500	0.18
		at 22°C	41	700	0.03

* Treatments for each catalyst done sequentially on same sample.
† 16 h reduction for Run 29; all other reductions for 1 h.

Changing the reduction temperature from 300 to 500°C for well dispersed catalysts (CAT-9, Runs 24 and 25; CAT-10, Runs 26 and 27) resulted in small decreases in hydrogen uptakes. These decreases in H/Pt ratios from about unity to 0.8 are probably due to sintering. The subtracted XRD pattern for Runs 26 and 27 in Figure 2B show a very slight maximum in the 40° region. Very little Pt growth is required to change Pt dispersion from unity to 0.8, and the XRD conditions used are not sensitive to Pt crystallites ≤ 2.0 nm. For a catalyst with a lower Pt dispersion (CAT-11), increasing the reduction temperature from 300 to 500°C did not change the H/Pt ratio appreciably. Hence, it is concluded that for $Pt/\gamma-Al_2O_3$ no anomalous hydrogen adsorption occurs for catalysts reduced at ≤ 500°C.

Increasing the reduction temperature to 700°C caused large decreases in hydrogen uptakes (see Runs 28,29,39 and 41, Table IV). Reduction at 700°C results in some Pt sintering as indicated by the maximum in Figure 2B for the subtracted patterns for Runs 29 minus 26, but this slight sintering cannot account for the decrease in H/Pt ratios from about unity to 0.19. A Pt/Al_2O_3 catalyst with a dispersion of 0.19 would have an intense Pt 111 line. This is illustrated in the top XRD patterns (Run 34) in Figure 2A. This XRD pattern was obtained for CAT-10 after the catalyst had been treated in oxygen at 700°C for 1 h. The H/Pt ratio for this sample after reduction at 500°C was 0.22, i.e. the H/Pt ratio for this case was higher than the one after reduction at 700°C. Hence, sintering is not the main reason for the low H/Pt ratios obtained after Runs 28 and 29.

The loss in hydrogen adsorption capacity which results from 700°C reduction can largely be restored by oxygen treatment at 22 to 100°C (cf. H/Pt ratios for Run 29 to Runs 30 and 31). The difference in H/Pt ratios for Run 27 and Runs 31 to 33 (i.e. 0.83 compared to 0.55) can be attributed to sintering which occurred during the 700°C reductions.

Many other investigators have observed significant decreases in hydrogen adsorption capacity as a result of high temperature reduction of Pt/Al_2O_3. These observations and possible explanations have been reviewed by Paal and Menon (12). Two possible causes which have been proposed are: one, the formation of a Pt-reduced Al_2O_3 surface complex (10-11), and two, the occurrence of very strong hydrogen adsorption and possibly absorption during high temperature reduction (18). The results in Table IV do not support the formation of Pt-reduced Al_2O_3 complexes since the suppression of hydrogen uptakes is less pronounced for catalysts with high Pt dispersions (CAT-10) than for catalysts with low Pt dispersions (CAT-11). Catalysts with high Pt dispersions should form surface complexes more readily than catalysts with low Pt dispersions since there is better Pt-support contact for highly dispersed catalysts, unless the presence of chlorine in CAT-10 suppresses surface complex formation under reducing conditions. However, the formation of a reduced Al_2O_3 complex which migrates onto the Pt particles and suppresses subsequent hydrogen adsorption cannot be ruled out.

The results in Table IV can also be explained in terms of strongly adsorbed/absorbed hydrogen which is not desorbed during subsequent degassing (18). Increasing the degassing temperature from 500°C (used for Run 41) to 700°C increased the H/Pt ratios from 0.03 to 0.06. This increase in H/Pt as a result of higher degassing

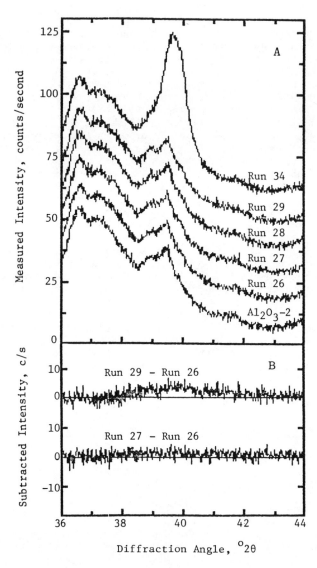

Figure 2. XRD patterns for CAT-10 (1.27% Pt/Al$_2$O$_3$-2).
A: Patterns as measured (offset for clarity);
B: Subtracted patterns (see Table IV for treatments).

temperature can also occur if the suppression in hydrogen adsorption
was due to poisoning or blocking of the Pt surface by a reduced Al_2O_3
species.

Platinum Supported on Silicalite. Hydrogen adsorption results in
Table V and XRD results in Figure 3 show that reduction at 500°C
after prior oxygen treatment at 550°C results in abnormally low H/Pt
ratios. The XRD results in Figure 3A show that oxygen treatment at
550°C causes significant sintering of the Pt (cf. XRD patterns for
Runs 42 and 43). However, reduction at 500°C does not cause
sintering (see subtracted patterns in Figure 3B).
    Hydrogen adsorption by fresh Pt/silicalite catalysts reduced at
500°C (CAT-12, Run 42 as well as for other Pt/silicalite catalysts we
have examined) appears to correspond to one hydrogen atom per surface
Pt atom. This conclusion is based on detailed comparisons of hydrogen adsorption results with XRD and electron microscopy results.
However, treatment in oxygen at $\geq$550°C followed by reduction at 500°C
results in hydrogen uptakes of less than one H atom per surface Pt
atom. For these oxygen treated Pt/silicalite catalysts the H/Pt
ratios obtained after 300°C reduction are a good measure of the Pt
dispersion.
    The low H/Pt ratios after 500°C reduction of the oxygen treated
catalysts have to be caused by poisoning of the Pt surface since
strong hydrogen adsorption/absorption does not occur during 500°C reduction. The nature of the poison is not known, but it could
originate from one of the precursors used in the preparation of
silicalite. It appears that oxygen treatment at elevated temperatures is required to mobilize this poison. The hydrogen adsorption
capacity can be restored by oxygen treatment at 400°C (see Runs 45 to
47), but subsequent reduction at 500°C again results in a low H/Pt
ratio (Run 48), i.e. the cycle of high and low hydrogen uptakes is
also reversible for Pt/silicalite.

Table V. Effect of Pretreatment on Hydrogen Adsorption for
0.98% Pt/Silicalite (CAT-12)

Oxygen Treatment*	Run	Reduction Temp. (°C)	H_2 Uptake (H/Pt)
none	42	500	0.58
1 h at 550°C	43	300	0.25
	44	500	0.14
1 h at 700°C	45	300	0.06
	46	500	0.00
1 h at 400°C	47	300	0.07
	48	500	0.01

* Treatments, reductions and adsorptions done sequentially
  on the same catalyst sample.

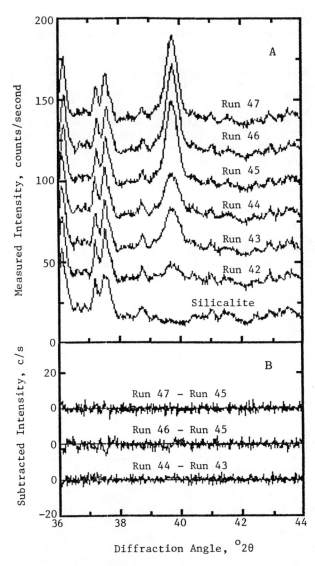

Figure 3.   XRD patterns for CAT-12 (0.98% Pt/Silicalite).
A: Patterns as measured (offset for clarity);
B: Subtracted patterns (see Table V for treatments).

Summary and Conclusions

The above results for Pt supported on three different types of difficult to reduce supports show that high temperature reduction usually causes large decreases in hydrogen adsorption capacity. However, the reduction temperature required for hydrogen adsorption suppression depends on the support. Hydrogen adsorption capacities of catalysts reduced at high temperatures can be restored by oxygen treatment followed by low temperature reduction, i.e. the processes responsible for adsorption suppression are reversible.

For all Pt supported on sulfur-containing MgO, 500°C reduction resulted in low hydrogen uptakes. This is attributed to sulfur poisoning of the Pt surface since for Pt on sulfur-free MgO 500°C reduction did not affect the hydrogen uptakes. The results for Pt/silicalite were similar to those for Pt on sulfur-containing MgO, but the decrease in hydrogen adsorption as a result of 500°C reduction was not as pronounced for the Pt/silicalite. Reduction at 500°C for Pt on alumina did not result in decreases in hydrogen adsorption capacities, but decreases were observed after 700°C reductions. The relative decrease in hydrogen adsorption for Pt/Al$_2$O$_3$ as a result of 700°C reduction appears to depend on Pt dispersion; catalysts with low Pt dispersion had relatively larger relative decreases in H/Pt than catalysts with high Pt dispersions. Further investigations are needed to determine whether the decreases in hydrogen adsorption for Pt/Al$_2$O$_3$ are caused by poisoning of a reduced form of alumina or by the formation of strongly adsorbed or absorbed hydrogen which is formed during high temperature reduction.

Acknowledgment

We acknowledge the support of this research by the Natural Sciences and Engineering Research Council of Canada.

Literature Cited

1. Spenadel, L.; Boudart, M. J. Phys. Chem. 1960, 64, 204–207.
2. Gruber, H.L. J. Phys. Chem. 1962, 66, 48–54.
3. Benson, J.E.; Boudart, M. J. Catal. 1965, 4, 704–710.
4. Mears, D.E.; Hansford, R.C. J. Catal. 1967, 9, 125–134.
5. Wilson, G.R.; Hall, W.K. J. Catal. 1970, 17, 190–206.
6. Hausen, A.; Gruber, H.L. J. Catal. 1971, 20, 97–105.
7. Wanke, S.E.; Dougharty, N.A. J. Catal. 1972, 24, 367–384.
8. Tauster, S.J.; Fung, S.C.; Garten, R.L. J. Am. Chem. Soc. 1978, 100, 170–175.
9. Tauster, S.J.; Fung, S.C. J. Catal. 1978, 55, 29–35.
10. Dautzenberg, F.M.; Wolters, H.B.M. J. Catal. 1978, 51, 26–39.
11. den Otter, G.J.; Dautzenberg, F.M. J. Catal. 1978, 53, 116–125.
12. Paal, Z.; Menon, P.G. Catal. Rev. 1983, 25, 229–324.
13. Wang, J.; Lercher, J.A.; Haller, G.L. J. Catal. 1984, 88, 18–25.
14. Flanigen, E.M.; Bennet, J.M.; Grose, R.W.; Cohen, J.P.; Patton, R.L.; Kirchner, R.M.; Smith, J.V. Nature 1978, 271, 512–516.
15. Margitfalvi, J.; Kern-Talas, E.; Szedlacsek, P. J. Catal. 1985, 92, 193–195.
16. Fiedorow, R.J.M.; Wanke, S.E. J. Catal. 1976, 43, 34–42.

17. Wanke, S.E.; Klengler, U.; Tesche, B. <u>Proc. 40th EMSA Mtg.</u> 1982, pp. 656-657.
18. Menon, P.G.; Froment, G.F. <u>Appl. Catal.</u> 1981, 1, 31-48.

RECEIVED September 12, 1985

18

Influence of Noble Metal Particles on Semiconducting and Insulating Oxide Materials

J. Schwank, A. G. Shastri, and J. Y. Lee

Department of Chemical Engineering, The University of Michigan, Ann Arbor, MI 48109-2136

In order to improve our understanding of metal-support interactions, metal-induced changes in structural properties of typical catalyst supports are investigated. Gold is chosen as metal component, and MgO and TiO$_2$ in form of anatase are used as support. In Au/MgO, a structural destabilization occurs which is a strong function of Au dispersion. Symptoms of the gold induced destabilization are: a significant weakening of the solid state oxygen bonds in MgO, increased susceptibility to electron beam damage in STEM, and Mg-carbonate formation following adsorption of CO at room temperature. In contrast to Au/MgO, the Au/TiO$_2$ system exhibits a mutual stabilization of the Au dispersion and of the high surface area anatase structure even after thermal treatment at high temperatures. No evidence for weakening of the oxygen bonds or carbonate formation is found. Our results show that, depending on the particular metal/oxide system, highly dispersed metal particles can cause significant modifications of the oxide properties. Consequently, blank studies which are generally carried out on catalyst supports to check for inertness in a given catalytic reaction may not always fully describe the behavior of these oxides after catalyst preparation.

The catalytic literature offers numerous examples for metal-support interactions, focusing mainly on support-induced changes in the adsorption characteristics and catalytic behavior of metals. Here, we are going to address the other side of this issue exploring certain manifestations of metal induced changes in the properties of typical oxide support materials. Since the pioneering work of Schwab, it is well known that the contact between a metal and a semiconductor can modify the catalytic activity of the semiconductor by altering the charge carrier density of the Schottky layers (1-3). The analogy between metal-semiconductor or metal-insulator contacts in electronic materials and heterogeneous catalysts containing small

0097-6156/86/0298-0182$06.00/0

metal particles supported on oxide materials is quite obvious. Besides electronic interactions it is conceivable that the presence of metal particles, especially when they are in a state of high or even atomic dispersion, might cause structural changes in the oxide materials, at least at or near the metal-oxide interface.

In the case of supported metal catalysts, the typical oxides used are generally high surface area materials which are prone to surface defects and poor crystallinity. The origin of the oxide material and the details of catalyst preparation and pretreatment will have a profound influence on the way how metal particles or small metal aggregates are in contact with the oxide surface. Numerous possible scenarios come to mind, all of them, of course, highly system specific. Without attempting to be comprehensive, examples include random 'adsorption' of metal atoms or metal aggregates on the oxide surface, grain boundary decoration, formation of two-dimensional metal rafts, deposition of metal particles of varying sizes and contact angles ranging all the way from wetting to non-wetting angles, formation of epitaxial overgrowth structures, formation of surface compounds, interstitial and substitutional solutions. To complicate matters even further, supported metal catalysts can undergo morphological and structural changes not only during catalyst pretreatment, but also under reaction conditions and during catalyst regeneration. To what extent such restructuring events come into the picture will depend on the properties of the metal-oxide interface. One might envision a situation where the interaction between metal and oxide leads to a mutual stabilization of both the metal dispersion and the oxide structure, or at the other extreme to a destabilization. Once again, analogies to material science come to mind where it is fully appreciated that the properties of materials can profoundly be influenced by the presence of impurities or dopants. In the case of supported metal catalysts having a metal loading of a few weight percent, one might view the metal as playing the role of an impurity. This aspect becomes particularly important when part of the metal gets incorporated or encapsulated in the oxide matrix.

The objectives of this work are to study the influence of gold particles on the properties of typical catalyst supports, namely MgO and TiO_2. Gold has been chosen because of its relatively low catalytic activity except for oxygen transfer reactions (4). MgO, an insulator, and TiO_2, a semiconducting material, are widely used as catalyst supports, and for both of them metal-support interactions have been reported in the literature. Our study places main emphasis on the role of gold on thermal stability, phase transformations, solid-phase oxygen exchange activity, and adsorption characteristics of the oxides.

Experimental

A multifaceted characterization effort to study these materials as a function of thermal treatment has been undertaken. The techniques include BET surface area measurements, X-ray diffraction, chemisorption, scanning and high-resolution transmission electron microscopy, analytical electron microscopy, neutron activation analysis, atomic absorption spectroscopy, FTIR and isotopic tracer studies. The details of catalyst preparation have been previously

reported (5-7). The Au catalysts were prepared by impregnation of
the support materials with aqueous solutions of $HAuCl_4$, followed by
reduction in either H_2 at 573-673 K or in oxalic acid at 623 K. Table
I gives characterization data, most of which were obtained in a
recent reinvestigation of the samples.

Table I. Catalyst Characterization Data

Catalyst Code	Au Loading (wt %)	Au Particle Size (nm)	Support	BET Area (m^2/g)	Cl Content (wt %)
32b	3.46	bimodal (1-10)	MgO (Carlo Erba)	248	0.02
1231	0.982	2.2	MgO (Carlo Erba)	264	0.04
1241	2.53	9.0	MgO (Fisher)	242	0.03
1241 sintered	2.53	47	MgO (Fisher)	35	–
1342	4.39	63	MgO (Carlo Erba)	31	–
Au/TiO$_2$	0.64	1.5	TiO$_2$ (Glidden)	118	0.01

The Carlo Erba MgO support was a low surface area oxide of about
15 m^2/g which underwent a massive increase in surface area upon
hydration, drying and subsequent reduction during the catalyst
preparation; residual chlorine contents exerted a significant
influence on the final surface area obtained (8). Table I shows the
surface areas obtained after catalyst reduction. The TiO$_2$ support,
supplied by Glidden, was obtained by hydrolysis of titanium
isopropoxide, followed by washing and drying at 378 K under vacuum.
The BET surface area of the freshly prepared TiO$_2$ was 118 m^2/g, and
appeared to be almost X-ray amorphous although high-resolution
transmission electron microscopy showed domains approximately 20 nm
in diameter with lattice fringes typical for anatase along with a few
domains which had the brookite structure (7). After heating for 2 hrs
to 673 K, X-ray diffraction showed patterns typical for pure anatase.
FTIR spectra were obtained with a Digilab FTS-20 Fourier Transform
infrared spectrometer. About 100 mg of catalyst was pressed into a
self-supporting wafer and placed into vacuum-tight infrared cells.
The catalyst wafers were pretreated inside the IR cell at 573 K in H_2
or O_2, followed by evacuation and cooling under vacuum to the desired
adsorption temperature. Gold particle sizes were determined by
oxygen chemisorption at 473 K, transmission electron microscopy, and
wide angle X-ray scattering. The details of the experimental
techniques have been reported earlier (6,7). Isotopic oxygen exchange

studies between gas phase and solid phase were carried out in a Pyrex-glass high-vacuum recycle reactor interfaced with a mass spectrometer ($\underline{9}$).

Results and Discussion

Au/MgO. In keeping with our objective to examine the influence of metal particles on the properties of oxides, first the question of possible modifications in the oxygen bond strength had to be addressed. An elegant way to assess the relative strength of oxygen bonds in solids is the measurement of the rates of isotopic oxygen exchange between solid phase and gas phase. Generally, heterogeneous oxygen exchange between oxides and labelled gas phase oxygen becomes significant only at elevated temperatures ($\underline{10,11}$). Since a previous study ($\underline{9}$) clearly showed that the presence of Au can trigger high isotopic oxygen exchange activity of MgO at temperatures where the blank oxide is almost inactive, it was of interest to explore more systematically the effect of gold dispersion and support surface area on the exchange activity.

Isotopic molecular oxygen can undergo three reactions in presence of a catalyst. The first reaction, the homogeneous exchange, is a scrambling of the oxygen isotopes which can take place either in the gas phase or on the catalyst surface without participation of oxygen from the solid phase.

$$^{32}O_2 + {}^{36}O_2 = 2\ {}^{34}O_2 \tag{1}$$

In our case, we start with an isotopic oxygen gas mixture which has already reached complete statistical equilibrium according to reaction (1). More interesting for our purposes is the participation of solid phase oxygen according to the following two heterogeneous exchange reactions:

$$^{36}O_2 + {}^{16}O_s = {}^{34}O_2 + {}^{18}O_s \tag{2}$$

$$^{36}O_2 + 2\ {}^{16}O_s = {}^{32}O_2 + 2\ {}^{18}O_s \tag{3}$$

Experimentally, the change in the relative amounts of molecular oxygen isotopes (mass 32, 34, and 36) and the total amount of ^{18}O in the gas phase are measured during the course of the reaction. Reaction (1) which is typically 2-3 orders of magnitude faster than the heterogeneous reactions (2) and (3) does not alter the content of ^{18}O in the gas phase. The overall exchange rate derived from the disappearance of ^{18}O from the gas phase combines in effect the rates of reactions (2) and (3). More details concerning the kinetic analysis are presented elsewhere ($\underline{9}$). Figure 1 shows that blank MgO and MgO containing large Au particles with an average particle size of 63 nm are practically inactive for uptake of isotopic oxygen at a temperature of 623 K. However, as the Au dispersion increases, the exchange activity becomes quite significant. Figure 2 gives the Arrhenius plots for Au/MgO catalysts of varying dispersion. The most active Au/MgO catalyst 32b has a bimodal particle size distribution with part of the Au in a state of very high, maybe even atomic dispersion.

As the gold dispersion decreases, the activity drops until it

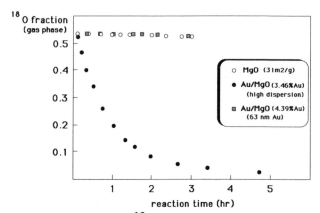

Figure 1. Concentration of ^{18}O in the gas phase as a function of reaction time in presence of MgO and two Au/MgO catalysts of high and low dispersion, respectively. The reaction was carried out in a recycle batch reactor at a total oxygen pressure of 2.13 kPa and at a temperature of 623 K.

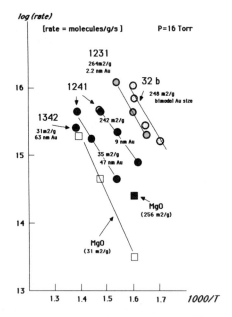

Figure 2. Arrhenius plots for the heterogeneous exchange of isotopic oxygen on MgO and Au/MgO catalysts of varying gold dispersion. See Table I for catalyst code numbers.

reaches that of blank MgO. Another important observation from Figure
2 is that the gold-induced increase in exchange activity by far
exceeds the change in activity one can achieve by increasing the BET
surface area of blank MgO. The exchange activity depends both on the
MgO surface area as well as the Au dispersion. However, Figure 2
clearly shows that the Au dispersion and probably the proportion of
Au in high (atomic?) dispersion is the dominating variable. These
conclusions are substantiated by a sintering experiment on the Au/MgO
catalyst 1241 where the agglomeration of Au from 9 to 47 nm
correlates nicely with a large drop in exchange activity. The amount
of oxygen atoms participating in the exchange reactions is
substantially higher than the total number of gold atoms in the
samples. This proves that oxygen from the support participates in
the reaction and that the exchangeable oxygen cannot be accommodated
on Au particles in form of gold oxides which are thermodynamically
unstable under reaction conditions.

One might wonder how gold could exert such a massive effect on
the isotopic oxygen exchange activity of MgO. A closer look at the
catalyst preparation conditions shows that the impregnation with
aqueous precursor solutions followed by drying at about 373 K can
lead to an almost complete bulk hydration of MgO to $Mg(OH)_2$.
Subsequent dehydration during catalyst pretreatment causes massive
changes in specific volume and surface area and can lead to the
formation of highly porous, high surface area MgO particles of poorly
defined crystal habit (8,12). The morphology of the resulting MgO
strongly depends on the thermal history and the impurity content of
the sample. It is conceivable that the nucleation and reduction of
gold takes place while these massive restructuring events are going
on in the $Mg(OH)_2$/MgO system. Some of the gold might get
incorporated or encapsulated by MgO, and some gold atoms might get
deposited on defect sites on the MgO surface thus destabilizing the
MgO oxygen bonds in the vicinity. This hypothesis is at least
qualitatively confirmed by electron microscopy results. Convergent
beam μ-μ electron diffraction patterns were obtained by focusing a
10nm diameter electron beam on chosen regions of the specimen. Blank
MgO gave μ-μ diffraction patterns which were reproducible if one
focussed the beam repeatedly on the same area for 5 minutes. However,
the μ-μ diffraction patterns of MgO taken in the immediate vicinity
of gold particles could only be obtained once. After about two
minutes of exposure to the electron beam, the diffraction patterns
tended to fade out indicating that localized regions of the MgO were
sensitive to electron beam damage. No such beam damage was
encountered in specimen regions which had no gold particles in the
immediate vicinity. The observations made on the JEOL-100 CX
transmission electron microscope operated at 100 kV were
independently confirmed by Dr. John Cowley at Arizona State
University (13) who investigated some of our MgO supported catalyst
samples in a 200 kV microscope.

To check whether the presence of gold would alter the adsorption
characteristics of MgO, an infrared study of CO adsorption at room
temperature was carried out. The results are shown in Figure 3. When
blank MgO that had been subjected to the same pretreatment as Au/MgO
except for the impregnation with the gold precursor salt was exposed
to 10 Torr of CO gas at 293 K, no adsorption of CO was observed and
the infrared spectrum showed no band whatsoever that could be

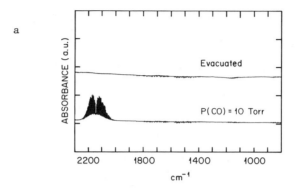

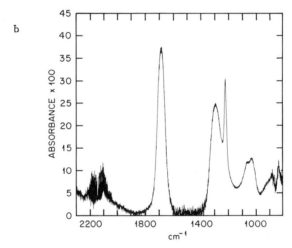

Figure 3 (a,b). Infrared spectra obtained after exposing
magnesium oxide and gold-magnesium oxide samples to CO gas
at a pressure of about 1.3 kPa. a) blank magnesium oxide;
b) highly dispersed 3.46% gold-magnesium oxide (catalyst
32b).

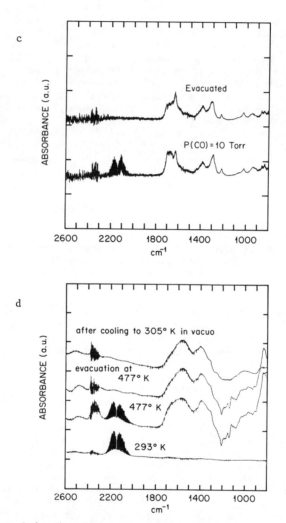

Figure 3 (c,d). Infrared spectra obtained after exposing magnesium oxide and gold-magnesium oxide samples to CO gas at a pressure of about 1.3 kPa. c) 4.39% Au-MgO (catalyst 1342); d) spectra obtained on blank MgO after heating in presence of CO and oxygen to 477 K.

attributed to adsorbed CO or to the formation of carbonates. In striking contrast to the lack of reactivity of the blank support towards CO, the Au/MgO catalysts showed strong infrared bands that could not be removed upon evacuation of the gas phase CO. While these bands fall into a spectral region that is typical for magnesium carbonates, the details of band position and relative band intensity depend on the individual catalyst sample. Here we show for comparison the spectra obtained on the catalysts representing the two extreme cases in the isotopic exchange experiment, namely the highly dispersed catalyst 32b and the Au/MgO catalyst 1342 with very low metal dispersion. Similar carbonate spectra although with slightly different band positions were also obtained on MgO catalyst supports impregnated with other metals such as ruthenium (14). In order to achieve carbonate formation on the blank MgO, it was necessary to heat the oxide in presence of a CO/O_2 gas mixture to 477 K. The resulting carbonate bands were quite different in appearance from those obtained after exposing the supported metal catalysts to pure CO at room temperature as can be seen from Figure 3. The fact that the carbonate bands are different depending on the nature of the metal and the metal dispersion could be taken as an indication that the carbonates are located at or near the metal–oxide interface. A more detailed band assignment and verification of this hypothesis will have to await further study. All these observations point towards a strong influence of Au on the properties of MgO, and we suspect that the proportion of Au in high dispersion plays a crucial role in destabilizing the oxide structure.

Au/TiO$_2$. A completely different picture emerges for the Au/TiO$_2$ system. There, our previous work (7) had already indicated that the presence of Au in high dispersion resulted in a mutual stabilization of the gold dispersion and the anatase structure of TiO_2. The latter was manifested by better thermal stability of the BET surface area of TiO_2 and by a significant delay in the anatase to rutile phase transformation. These results were interpreted in terms of gold forming an interstitial solution and decorating the TiO_2 grain boundaries, thus restricting the cooperative movement of Ti^{4+} and O^{2-} ions which would be required to accomplish the phase transformation. In keeping with this idea of gold-induced stabilization of TiO_2 one would not expect gold to destabilize the oxygen bonds in the solid phase in a manner analogous to that seen on MgO. In fact, our highly dispersed Au/TiO$_2$ catalyst showed no indication of any enhanced isotopic oxygen exchange activity as compared to blank TiO_2.

In contrast to MgO, exposing the Au/TiO$_2$ catalyst to CO at 293 K and 323 K did not lead to the appearance of any carbonate bands. The adsorption isotherms obtained on blank TiO_2 and Au/TiO$_2$ at 293 K showed considerable uptake of CO on the blank support while at 323 K the CO uptake on the support became almost negligible (7). At both temperatures, most of the adsorbed CO could easily be removed by prolonged evacuation, indicating low heats of adsorption on both Au and TiO_2 surfaces. CO adsorption on Au typically gives a single infrared band at about 2110–2070 cm^{-1} (15–21). To the best of our knowledge, there has been no previous infrared study of CO adsorption on Au/TiO$_2$. According to Yates (22), CO adsorbed on TiO$_2$ gives two infrared bands at 2180 and 2210 cm^{-1}. A recent publication (23) reports three infrared bands of CO on the anatase form of TiO_2 at 2110, 2190, and 2210 cm^{-1}. The band at 2110 cm^{-1} was assigned to CO adsorption to reduced Ti^{3+} sites while the bands at higher wavenumbers were attributed to CO adsorption on weak and strong Lewis acid sites, respectively. These Lewis acid sites are believed to be Ti^{4+} cations with different degrees of coordinative unsaturation which are exposed at the oxide surface.

In our case, we found that on high surface area anatase three CO bands appeared at 2128, 2188, and 2210 cm^{-1} after room temperature adsorption of CO at a pressure of 10 Torr (Figure 4). In agreement with reference ([23]) we assign the three bands also to adsorption on Ti^{3+}, and week and strong Lewis acid sites on the TiO_2 surface. The CO band at 2128 cm^{-1} took longer to remove by evacuation compared to the two other bands indicating a higher strength of adsorption of CO on Ti^{3+} sites. The two bands at higher wavenumbers which are attributed to CO adsorption on Lewis acid sites appeared to be rather sensitive to catalyst pretreatment probably leading to slight variations in hydroxyl group concentrations on the TiO_2 surface. The infrared spectra of CO on Au/TiO_2 are also shown in Figure 4. All three bands that were seen on the blank TiO_2 are visible. The low wavenumber band at 2128 cm^{-1} is now clearly the most intense band in the spectrum and appears to be slightly enhanced compared to the blank TiO_2 spectrum. With decreasing relative CO surface coverage, the two bands at higher wavenumbers decrease in intensity without a shift in band position. The low wavenumber band, once again, is more stable. The position of the band maximum shifts from 2128 to 2125 cm^{-1} as the surface coverage decreases, a shift similar to that observed on blank TiO_2. The question now arises whether the increased band intensity of the 2128 cm^{-1} band is due to adsorption of CO on Au sites which should give a band in the region of 2120 to 2070 cm^{-1}. The band position of 2128 cm^{-1} obtained on Au/TiO_2 is higher than any band position ever reported for CO on Au. Furthermore, it has been observed that the infrared band of CO adsorbed on Au shifts to higher wavenumbers as the surface coverage decreases ([21]), a trend opposite to the one observed here. Finally, CO adsorbed on Au can usually be removed just by brief evacuation. Here, the low wavenumber band proved to be quite persistent and required prolonged evacuation before it could be removed. All these facts indicate that the 2128 cm^{-1} band could be a result of an interaction between highly dispersed Au located in interstitial positions and on grain boundaries of the oxide and the titanium cations resulting in an increased concentration of Ti^{3+} sites.

The uptake of CO by the TiO_2 support became almost negligible when the adsorption temperature was raised to 323 K, while the Au/TiO_2 catalyst was still able to adsorb CO at this temperature ([7]). The infrared spectra of CO adsorbed at 323 K on blank anatase and on Au/TiO_2 are shown in Figure 4c and 4d. In contrast to the room temperature experiment, CO adsorption at 323 K led to only one weak band at about 2128 cm^{-1}. Once again, no band attributable to CO adsorbed on Au could be detected. It seems that gold in high dispersion can contribute to a slight enhancement of CO adsorption on anatase. Some of the CO might be adsorbed at the metal-oxide interface mainly in the form of CO on Ti^{3+} cations.

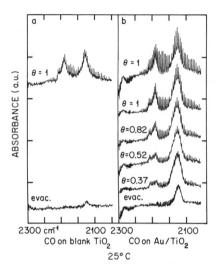

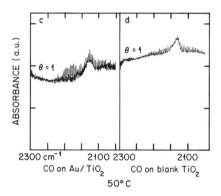

Figure 4. Infrared spectra of CO on a) titanium dioxide at a) 298 K and b) 323 K; and gold-titanium dioxide at c) 298 K and d) 323 K. The values of theta indicate the relative residual CO pressure in the IR cell normalized with respect to a CO partial pressure of 1.3 kPa.

Conclusions

Our study has provided some interesting clues with regard to metal-induced changes in typical catalyst support materials. It is obvious that the phenomena observed are system specific, and our data do not permit us to extrapolate our findings in a straightforward manner to other supported catalyst systems. It is also not clear whether or not the enhancement of the carbonate formation or the increased solid-phase oxygen mobility in MgO is related to support effects reported for catalytic reactions on MgO supported metals. Nevertheless, it seems important to keep a wary eye on metal-induced changes in typical catalyst supports. Even if blank experiments prove that there is no activity of the support under given reaction conditions, one should not tacitly assume that these results apply to the support after it has been impregnated with metal.

Acknowledgments. The authors would like to thank Dr. S. Galvagno for his contribution to catalyst preparation and characterization. Acknowledgment is made to the Donors of the Petroleum Research Fund administered by the American Chemical Society for partial support of this research. Support through ARO is also gratefully acknowledged.

References

1. Schwab, G. M. <u>Trans. Faraday Soc</u>. 1946, 42, 689.

2. Schwab, G. M. <u>Discuss. Faraday Soc</u>. 1950, 8, 166.

3. Schwab, G. M. <u>Naturwiss</u>. 1957, 44, 32.

4. Schwank, J. <u>Gold Bulletin</u> 1983, 16, 103.

5. Galvagno, S.; Parravano, G. <u>J. Catal</u>. 1978, 55, 178.

6. Fukushima, T.; Galvagno, S.; Parravano, G. <u>J. Catal</u>. 1979, 57, 177.

7. Shastri, A. G.; Datye, A. K.; Schwank, J. <u>J. Catal</u>. 1984, 87, 265.

8. Leofanti, G.; Solari, M.; Tauszik, G. R.; Garbassi, F.; Galvagno, S.; Schwank, J. <u>Appl. Catal</u>. 1982, 3, 131.

9. Schwank, J.; Galvagno, S.; Parravano, G. <u>J. Catal</u>. 1980, 63, 415.

10. Winter, E. R. S. <u>J. Chem. Soc. Ser. A</u> 1968, 12, 2889.

11. Boreskov, G. K. <u>Advan. Catal</u>. 1964, 15, 285.

12. Shastri, A. G.; Chae, H. B.; Bretz, M.; Schwank, J. <u>J. Phys. Chem</u>. 1985 (in press).

13. Cowley, J. private communication.

14. Lee, J. Y. Ph.D. Thesis, The University of Michigan, Ann Arbor, 1985.

15. McElhiney, G.; Pritchard, J. Surface Sci. 1976, 60, 397.

16. Guerra, C. R.; Schulman, J. H. Surface Sci. 1967, 7, 229.

17. Bradshaw, A. M.; Pritchard, J. Proc. Roy. Soc. (London) 1970, A316, 169.

18. Kottke, M. L.; Greenler, R. G.; Tompkins, H. G. Surface Sci. 1972, 32, 231.

19. Yates, D. J. C. J. Colloid Interface Sci. 1969, 29, 194.

20. Guerra, C. R. J. Colloid Interface Sci. 1969, 29, 229.

21. Schwank, J.; Parravano, G.; Gruber, H. L. J. Catal. 1980, 61, 19.

22. Yates, D. J. C. J. Phys. Chem. 1961, 65, 746.

23. Busca, G.; Saussey, H.; Saur, O.; Lavalley, J. C.; Lorenzelli, V. Appl. Catal. 1985, 14, 245.

RECEIVED September 12, 1985

Evidence of a Metal–Surface Phase Oxide Interaction for Re on WO$_x$ Supported on Activated Carbon

L. L. Murrell, N. C. Dispenziere, Jr., R. T. K. Baker, and J. J. Chludzinski

Corporate Research Science Laboratories, Exxon Research and Engineering Company, Annandale, NJ 08801

Hydrogen and CO chemisorption of Re, W, and Re-W on a carbon black support reduced in H$_2$ confirm metallic phases are present in all three systems. The temperatures at which reduction occurs, as well as an apparent strong metal support interaction between Re and supported WO$_x$, was followed by chemisorption. In addition, the catalytic gasification of the carbon black by the very active Re component in H$_2$ was significantly retarded for the Re-W system. Controlled atmosphere electron microscopy examination of the Re-W system confirmed the marked decrease in the catalyzed gasification of graphite in H$_2$ compared to the very active Re component. In the gasification of graphite in O$_2$, however, the Re-W bicomponent alloy system was significantly more active than either tungsten or rhenium. It was concluded that at high temperatures in O$_2$, Re covers the Re-W alloy particles, and at high temperatures in H$_2$, W covers the Re-W alloy particles. At low temperatures highly-dispersed Re particles interact with a WO$_x$ surface phase complex either on the carbon surface or on the Re particle surface to retard both H$_2$ and CO chemisorption.

There is considerable current interest in the strong interaction between certain oxide supports and a metal phase (SMSI) (1,2). In addition, the interaction between oxide supports and highly dispersed oxide phases is receiving increasing attention (3-6). This paper presents evidence that a relatively inert activated carbon black support, (Carbolac, containing 20 wt% oxygen, Cabot Corporation), when used as a support for a transition metal oxide (TMO) alters the properties of the supported TMO in significant ways. Bulk tungsten oxide does not behave as an SMSI support whereas tungsten oxide when supported on activated carbon is an SMSI support. The extent of reduction of TiO$_2$ has been implicated in the SMSI phenomenon (7,8). In this paper we present clear evidence that the number of systems which can give rise to the SMSI phenomenon is not limited to primary oxides such as TiO$_2$, Nb$_2$O$_5$, or binary oxides such as BaTiO$_3$ or ZrTiO$_4$.

0097-6156/86/0298-0195$06.00/0

In a related study of Rh on TiO_2-doped SiO_2 (9) it was shown that a highly-dispersed surface phase complex of TiO_2 on an inert SiO_2 support introduced coordinatively unsaturated sites (CUS) (10-12) of Ti^{+4} which were "titrated" by the reduced metal clusters. This Rh-TiO_2 on SiO_2 interaction lead to CO and H_2 chemisorption suppression when reduced in H_2 at 500°C. Similar studies of FeO on TiO_2 and Ir-Fe on TiO_2 (13) show interaction of the CUS sites of TiO_2 with both Fe^{+2} centers and an Ir-Fe bimetallic cluster. Activated carbon in the present study and graphite model systems have been employed as supports for WO_3, Re_2O_7 and WO_3-Re_2O_7. The properties of these three systems in reducing and oxidizing environments have been studied by H_2 and CO chemisorption, x-ray diffraction (14), and Controlled Atmosphere Electron Microscopy (15). Rhenium, W, and Re-W are interesting systems to investigate as far as the graphite gasification in oxidizing and reducing conditions is concerned (16,17). This is due to the fact that the single component systems Re and W, in both oxidizing and reducing environments show such extremes in catalytic activity. Tungsten is completely inert toward graphite gasification in H_2 whereas rhenium is one of the most active catalysts (15). Tungsten is also only marginally more active than the uncatalyzed graphite gasification in O_2 whereas rhenium is an active catalyst. Studies of the mixed component system, Re-W, are therefore of considerable interest (16,17) since the single component systems have such different catalytic activity toward graphite gasification.

The Re-W system shows complete reversal in gasification activity for oxidizing and reducing conditions. The Re-W system for graphite gasification in oxygen is more active than rhenium itself. This is consistent with a model wherein a particle of W is surrounded by an outer Re metal phase. This "cherry model" of the Re-W system explains enhanced activity for graphite oxidation either via the thin "outer skin" of Re allowing faster carbon diffusion and/or via the Re phase being altered electronically by the inner W core. In any case the Re-W gasification in oxygen is clearly more characteristic of Re than W, which is nearly inert.

The Re-W system for graphite gasification in hydrogen is just the opposite to gasification in oxygen. Rhenium-tungsten is completely inert for graphite gasification in hydrogen! This result argues that under reducing conditions the outer surface of the Re-W metal particles are covered by an "outer skin" of W. The W surface segregation completely retards the activity of one of the most active graphite hydro-gasification catalysts known. The CAEM studies of Re, W and Re-W on graphite are only one aspect of their interesting chemistry. Chemisorption and x-ray diffraction studies of the Re, W and Re-W systems shows that WO_x on activated carbon to be SMSI support (1,2). The H_2 and CO chemisorption uptake for W on activated samples reduced at 500°C in flowing H_2 was very low despite W metal particles being detected by x-ray diffraction, see Tables 1 and 2. Since bulk WO_3 does not reduce until 550°C we suspected oxygen contamination of the W metal particle surface. Reduction of both a 10 and a 25% W on activated carbon sample at 700°C increased both the H_2 and CO chemisorption to that expected of a dispersed metal particle of ca 10 nm diameter, see Tables 1 and 2.

Table I. Hydrogen Chemisorption and X-ray Diffraction Studies of W, Re, and Re-W on Carbolac

Sample on Carbolac	Reduction Temperature(°C)	Metal Surface Area[1] by H_2 Chemisorption(m^2/g)	Metal Surface Area Measured by X-ray Line Broadening(m^2/g)
10% W/C	500	0.53	Very Broad Line
10% W/C	700	15.9	43
25% W/C	500	2.61	50
25% W/C	700	10.8	28
10% Re/C	300	91.5	NV[2]
10% Re/C	500	37.9	23
10% Re-10% W/C	300	13.6	NV[2]
10% Re-10% W/C	500	9.88	~60

[1] The H_2 chemisorption blank on Carbolac was found to be 0.056 cc/g reduced at 300° or 500°C and 0.067 cc/g reduced at 700°C. This small gas uptake on the Carbolac support was subtracted from the values for all the samples in this table.

[2] Not visible.

In contrast to the sluggish reduction of W on activated carbon 10% Re on activated carbon reduced readily at 300°C as confirmed by H_2 and CO chemisorption. This fact will become important when comparison is made to Re-W on activated carbon. Reduction of Re on activated carbon in flowing H_2 at 500°C for 1 hr. resulted in about 50% gasification of the activated carbon. Considerable sintering of the 500°C reduced sample occured compared to the 300°C reduced sample as confirmed both by x-ray diffraction and by H_2 and CO chemisorption. Reduction of a sample of 10% Re on activated carbon at 700°C in flowing H_2 resulted in nearly complete carbon gasification in 5 min. The high carbon gasification activity of Re on graphite is consistent with the observed gasification of activated carbon.

A 10 wt.% Re-10 wt.% W on activated carbon sample was next investigated by H_2 and CO chemisorption. The sample was prepared by first introducing the W component followed by introduction of the Re component. The H_2 and CO chemisorption of the Re-W on activated carbon reduced at 300°C was reduced by a factor of 10 compared to Re on activated carbon. Trivial sintering of Re to give large particles was discounted as x-ray diffraction showed that no Re particle could be detected. Increasing the reduction temperature to 500°C for Re-W on carbon failed to increase the H_2 and CO chemisorption uptake. Also, the carbon gasification observed for the 10% Re on

activated carbon sample was completely suppressed in the case of Re-
W on activated carbon. These observation argues that Re interacts
with the highly dispersed WO_x phase on activated carbon at both 300
and 500°C in H_2. This interaction alters both chemisorption and
carbon gasification characteristic of the Re component. The model

Table II. Carbon Monoxide Chemisorption and X-ray Diffraction
Studies of W, Re, and Re-W on Carbolac

Sample on Carbolac	Reduction Temperature(°C)	Metal Surface Area[1] by H_2 Chemisorption(m^2/g)	Metal Surface Area Measured by X-ray Line Broadening(m^2/g)
10% W/C	500	2.86	Very Broad Line
10% W/C	700	13.1	43
25% W/C	500	6.29	50
25% W/C	700	17.4	28
10% Re/C	300	123	NV^2
10% Re/C	500	19.0	23
10% Re-10% W/C	300	21.2	NV^2
10% Re-10% W/C	500	12.7	60

[1] The CO chemisorption blank on Carbolac was found to be 0.09 cc/g
when reduced at 500°C and zero when reduced at 700°C. The uptake
on the Carbolac support was subtracted from the values for all the
samples described in this table employing the double isotherm
technique.

that partial reduction of transition metal oxide phases leads to the
SMSI phenomenon must be expanded to a dispersed transition metal
oxide phase on a carbon support. CAEM studies support the view that
WO_x on carbon is a highly dispersed and strongly interacting phase
with the carbon substrate. A recent paper (18) describing Pt-Mo
bimetallic catalysts supported on X-zeolites reports a significant
decrease in H_2 and CO chemisorption as Mo interacts with the surface
of the 1.0 nm Pt particles in the zeolite cages. The parallels
among the results for Pt-Mo on X-zeolite, the results for Re-W on
activated carbon in this paper, and the results for SMSI systems on
TiO_2 (1,2,7,8,13) are striking.

References

1. Tauster, S. J., Fung, S. C. and Garten, R. L.; J. Am. Chem. Soc. 1978, 100, 170.
2. Tauster, S. J. and Fung, S. C.; J. Catal. 1978, 54, 29.
3. Lund, C. R. F. and Dumesic, J. A.; J. Phys. Chem. 1981, 85, 3175.
4. Yermakov, Yu. I., Kuznetsov, B. N. and Zakharov, V. A., "Catalysis by Supported Complexes"; Elsevier, 1981.
5. Yuen, S., Chen, Y., Kubsh, J. E. and Dumesic, J. A.; J. Phys. Chem. 1982, 86, 3022.
6. Soled, S., Murrell, L., Wachs, I. and McVicker, G.; Am. Chem. Soc. Div. Pet. Chem. Prepr. 1983, 28, 1310.
7. Baker, R. T. K., Prestridge, E. B. and Garten, R. L.; J. Catal. 1979, 59, 293.
8. Baker, R. T. K., Prestridge, E. B. and Murrell, L. L.; J. Catal. 1983, 79, 348.
9. Murrell, L. L. and Yates, D. J. C.; Stud. Surf. Sci. Catal. 1981, 7, 1470.
10. Basset, J. M. and Ugo, R., In R. Ugo (Ed.); "Aspects of Homogeneous Catalysis", Reidel: Dordecht, 1976; Vol. III, p. 170.
11. Anderson, J. R., Elmes, D. S., Howe, R. F. and Mainwaring, D. E.; J. Catal. 1977, 50, 508.
12. Peri, J. B.; J. Phys. Chem. 1982, 86, 1615.
13. Murrell, L. L. and Garten, R. L.; Applications of Surface Science, (in press).
14. Murrell, L. L. and Dispenziere, N. C. Jr., (submitted to J. Applied Catalysis).
15. Baker, R. T. K., Chludzinski, J. J. Jr., Dispenziere, N. C. Jr. and Murrell, L. L.; Carbon, 1983, 21, 579.
16. Baker, R. T. K., Sherwood, R. D. and Dumesic, J. A.; J. Catal. 1980, 62, 221.
17. Baker, R. T. K., Sherwood, R. D. and Dumesic, J. A.; J. Catal. 1980, 66, 56.
18. Tri, T. M., Cordy, J-P., Gallezot, P., Massandier J., Primet, M., Vedrine, J. C. and Imelik, B.; J. Catal. 1983, 79, 396.

RECEIVED September 17, 1985

20

Metal–TiO$_2$ Catalysts

Electronic Effects During H$_2$ Chemisorption, CO–H$_2$ Interactions, and Photocatalysis

J.-M. Herrmann

Equipe photocatalyse, C.N.R.S., Ecole Centrale de Lyon, B.P. 163, 69131 Ecully Cedex, France

The electronic behaviour of Pt, Rh and Ni/TiO$_2$ catalysts was followed by measuring in situ the conductivity σ of the support. From σ variations at room T, it is deduced that the metal is always enriched in free electrons from the support and that H$_2$ chemisorption is followed by H spill over on the oxide. On Pt/TiO$_2$, σ measurements showed (i) that CO chemisorbs as a donor molecule, which can explain why an excess of electrons in the metal can counteract CO chemisorption and (ii) that CO does not dissociate on TiO$_2$ or at the interface. At reaction temperature (290°), CO does not dissociate either, which can hint that methanol is a primary product and, on adding H$_2$, the decrease of σ shows that TiO$_2$ surface is reoxidized by the oxygenated products (methanol, water), thus partly destroying SMSI. The electron exchange between the metal and its support play a fundamental role in SMSI as clearly demonstrated in the photocatalytic cyclopentane-deuterium isotopic exchange reaction.

Recently titania appeared as a non-conventional support for noble metal catalysts, since it was found to induce perturbations in their H$_2$ or CO adsorption capacities as well as in their catalytic activities. This phenomenon, discovered by the EXXON group, was denoted "Strong Metal-Support Interactions" (SMSI effect) (1) and later extended to other reducible oxide supports (2). Two symposia were devoted to SMSI, one in Lyon-Ecully (1982) (3) and the present one in Miami (1985) (4) and presently, two main explanations are generally proposed to account for SMSI: (i) either the occurence of an electronic effect (2,5-13) or (ii) the migration of suboxide species on the metal particles (14-20). The second hypothesis was essentially illustrated on model catalysts with spectroscopic techniques. It can be noted that both possibilities do not necessarily exclude each other and can be considered simultaneously (21).

In the present paper, the metal-support electronic interactions in various metal catalysts—mainly Pt – were followed by measuring in situ the electrical conductivity of the solids either in the "normal" or the "SMSI" state, when in contact with various atmospheres (vacuum, H$_2$, O$_2$, (CO+H$_2$). The manifestation of the electronic factor

0097-6156/86/0298-0200$06.00/0

during catalysis was illustrated by the photocatalytic isotopic exchange between deuterium and cyclopentane.

Experimental

Catalysts:The three types of catalysts (Pt, Rh, Ni/TiO$_2$) where
prepared by impregnating a desired quantity of titanium dioxide
(anatase Degussa P-25; 50 m^2 g^{-1}) with aqueous solutions of compounds containing the cations of the metal chosen (chloroplatinic
acid, rhodium trichloride and nickel hexammine nitrate) at concentrations appropriate to yield the proper metal loading (5 wt% unless
otherwise stated). The impregnated slurries were subsequently evacuated at 80°C in a rotating flask, dried at 110°C for 2 h, reduced
at 480°C overnight in hydrogen, cooled down to room temperature under nitrogen flow and kept in vials untill further use. Chemical
analyses gave metal loadings of ≃ 5 wt%.
    The catalysts were characterized by hydrogen chemisorption and
transmission electron microscopy (TEM). Pt and Rh were present as
homodispersed particles of 2 and 3.5 nm respectively, homogeneously
distributed on all the quasi-spherical (d∿25 nm) particles of P-25
Degussa anatase. Concerning Pt, different catalysts were prepared
with loadings varying from 0.1 to 10%: the particle size remained
unchanged (1.5-2 nm) even with a metal loading varying by 2 orders
of magnitude (22).
    Nickel particles were far bigger and because of a lack of con-
trast could not be seen by TEM. Magnetic measurements, however,
yielded a mean size of about 13.5 nm. A rough estimation showed that
only 10 to 20% of TiO$_2$ particles were in contact with a nickel one.

Electrical conductivity measurements: They were carried out in
a static-type cell, designed for powder samples, which allows in
situ measurements from the beginning of the pretreatment up to subsequent solid-gas interactions. The results refer only to the support in the case of supported metal catalysts.
    The photocatalytic isotopic exchange between cyclopentane and
deuterium (CDIE) was carried out in a static fused silica photoreac-
tor described in ref (23).

Results and Discussion

Effect of SMSI on H$_2$ chemisorption: The various M/TiO$_2$ catalysts
were reduced in 250 Torr H$_2$ either at low temperature (200°C; LTR
samples) or at high temperature (500°C; HTR samples). LTR Ni/TiO$_2$
sample was exceptionnally reduced at 300°C to make sure that all
nickel was completely reduced. H$_2$ adsorption capacities are listed
in Table I.
    From Table I, it can be seen that for a constant metal loading
(5%), the sensitivity to SMSI of the present samples varies as
Ni ≫ Pt>Rh. This classification is purely qualitative and not indicative of the proper influence of the nature the metal since several
parameters such as texture and dispersion can influence the extent
of SMSI. For instance, nickel is under the shape of far larger particles than Pt or Rh ones with 2 to 3 times more atoms. On the contrary, in the case of Pt catalysts, the comparison between catalysts
is more meaningful since the metal crystallite size is homodispersed

and constant. It can be observed from Table I (i) that for two different preparations with the same Pt content (5%), the reproducibility is correct and (ii) that for two different loadings (0.5 and 5 Pt%), the effect of SMSI on the adsorption is stronger for the smaller one and initiates at lower temperatures (300°C). This behaviour will be observed and discussed further in the catalytic section.

Table I. Adsorption of H_2 on various M/TiO_2 catalysts

Catalysts		Amount of chemisorbed H_2 ($\mu mol/g.cat$)		
		Reduction temperature		
		200	300	500
Pt/TiO_2	5%	41	41	6
Pt/TiO_2	5%	38	—	4
Pt/TiO_2	0.5%	6.5	3	0
Rh/TiO_2	5%	67	—	24
Ni/TiO_2	5%	—	8	0

Electrical conductivity study of M/TiO_2 catalysts

This study has been previously described in ref (11,24) and can be summarized as follows.

(i) LTR samples ($T_R = 200°C$): In H_2 at 200°C, since the presence of a metal deposited increases the conductivity of titania by 1 to 2 orders of magnitude, it was inferred that the metal catalyzes the reduction of the oxide with spilt over atomic hydrogen which first creates OH_S^- surface groups:

$$H_2 + 2\ M_S \rightleftharpoons 2\ M_S - H \qquad (1)$$

$$M_S - H + O_S^{2-} \rightleftharpoons M_S + OH_S^- + e^- \qquad (2)$$

whose dehydration forms anionic vacancies V_O^{2-}

$$2\ OH_S^- \rightleftharpoons H_2O(g) + O^{2-} + V_O^{2-} \qquad (3)$$

which are generally singly ionized at the temperatures considered

$$V_O^{2-} \rightleftharpoons V_O^+_{2-} + e^- \qquad (4)$$

Simultaneously, because of the good physical and thence electrical contact between both solid phases, there is an alignment of their Fermi levels which demands an electron migration - even limited - from the support to the metal in agreement with the respective values of the work functions of the metal and of the reduced support.

$$e^- + M \rightleftharpoons e_M^- \qquad (5)$$

The electron transfer of Eq (5) is confirmed when H_2 is evacuated at 400°C since the electrical conductivity of M/TiO_2, $\sigma_{M/TiO2}$, decreases whereas σ_{TiO2} increases because of the creation of new anionic vacancies due to the increase of temperature.

$$O_S^{2-} \rightleftharpoons 1/2\ O_2(g) + V_O^+_{2-} + e^- \qquad (6)$$

In Eq(6), the charge balance is obtained by summation in both sides of the equation of all the charges mentioned, included those in subscripts (like $V_{\ddot{O}}^{2-}$) in order to account for the fact that species such as O^{2-} and V_O^{2-} are neutral with respect to the solid. Eq(5)is also confirmed when T is decreased under vacuum from 400°C to room temperature since M/TiO_2 samples remain semiconductors with a constant activation energy of conduction E_C, in contrast with bare titania which behaves as a quasi-metallic conductor ($E_C \simeq 0$).

Finally, the introduction of hydrogen at room temperature on M/TiO_2 produces a reversible increase of σ_{M/TiO_2} which follows the isotherm law:

$$\sigma = a + b \, P_{H_2}^{1/2} \qquad (7)$$

which can be accounted for by Eqs (1), (2) and (5).

(ii) HTR samples (T_R = 500°C), SMSI state: Qualitatively, M/TiO_2 samples behave as the LTR ones. However, as evidenced by the higher value of σ, the reduction of titania is much more important. A Ti_4O_7 phase was even identified (18,25). Consequently, the relative electron enrichment of the metal is stronger in agreement with the alignment of the Fermi levels of the metal and of reduced titania whose work function decreases with the reduction level. This electron excess in the metal, even if it is limited, is thought to be at the origin of the SMSI effect which partly suppresses H_2 chemisorption, especially if this chemisorption occurs with the creation of dipoles at the surface of metal (26,27,28). The absence of chemisorptive properties under SMSI conditions for catalysts having a low loading, which depends on the nature and texture of the metal (0,5% for Pt and 5% for Ni, Table I) is confirmed by the absence of hydrogen spill over ($d\sigma/dP_{H_2}$ = 0 in Eq.(7))on these samples, whereas for all other samples whose H_2 chemisorption capacity is not nil when in the SMSI state, the relationship of Eq 7 is duly observed.

The restauration of a normal state by exposure to oxygen is explained by the reoxydation of the support with the filling of surface anionic vacancies (Eq.–6) which increases titania's work function and requires the retrocession of excess electrons to the oxide.

$CO + H_2$ interactions on 5 wt% Pt/TiO_2 in the SMSI state

Interactions at room temperature: When CO is first introduced (Fig.1), σ increases instantaneously and then remains independent of P_{CO}. The fact that σ does not decrease means that CO does not dissociate on titania nor at the interface, otherwise the filling of anionic vacancies by atomic oxygen (Eq.–6) would have decreased substantially σ by consuming free electrons. The sharp initial increase, on the contrary, shows that CO chemisorb on Pt with a donor effect probably due to the creation of dipoles as proposed for H_2 chemisorption which renders ohmic the electrical contact between the metal and its semiconductor support (26, 17, 28).According to these authors, the creation of a dipole layer decreases the work function $\emptyset_M$ of the metal which approaches the electron affinity of the semiconductor, thus suppressing the Schottky barrier. Presently CO adsorbs as a donor molecule on Pt decreasing $\emptyset_{Pt}$, which allows electrons to be restituted to titania. The absence of variations of σ versus P_{CO} for subsequent increasing CO pressure means that the surface is already saturated at these pressures.

The introduction of H_2 in the presence of CO increases σ but more slowly than did the initial dose of CO. The kinetics is comparable to that of hydrogen spillover which can be observed on this

catalyst since H_2 chemisorption under SMSI conditions is not nil
(Table I). This means that H_2 is able to displace CO from certain
of its sites (29) from which it can subsequently migrate on the surface of titania. This requires that these adsorption sites are the
weakest for CO and close to the interface to let hydrogen spill over.

Interactions at reaction temperature (290°C): When the same experiments are repeated at 290°C (Fig.2), CO does not dissociate on
titania since σ does not decrease but on the contrary conserves its
donor character to platinum which retrodonates some excess electrons
to the support.

The introduction of H_2 increases first sharply σ for a few seconds and then produces a rather slow decrease. This behaviour shows
that H_2 is always-able to adsorb dissociatively on a surface saturated with CO but now the spill over is prevented – or at least masked – by the reaction itself whose oxygen – containing products,
mainly methanol and water, are able to partially reoxidize the surface of titania by filling some of its anionic vacancies. A similar
decrease of σ can be obtained by introducing pure water. The sharp
initial increase of σ observed in Fig.2 is to be ascribed to a thermal effect due to the starting reaction and the subsequent exothermic reoxidation of titania.

The drop of σ induced by the partial reoxidation of titania's
surface is small in comparison with the reoxidation by gaseous oxygen at room temperature (11,24).This is not unexpected since (i)
oxygen is highly electrophilic and readily chemisorbs on reduced
titania; (ii) the temperature of reoxidation by O_2 is by 290°C lower
than for (CO + H_2), which does not unfavor the strongly exothermic
reoxydation of TiO_2 but substantially decreases σ because TiO_2 has
recovered its complete semiconductor behaviour with a high activation energy of conduction and (iii) in the case of (CO + H_2), the
atmosphere remains highly reducing, permitting the maintenance of
anionic vacancies with free electrons of conduction. In fact, the
σ level at the end of (CO + H_2) interactions at 290°C is comparable
to that of Pt/TiO_2 under H_2 at 200°C in the "normal state".

Consequently, although CO and H_2 chemisorptions are decreased
by a factor of 5 and 9 respectively on the 5% Pt sample used, the
absence of inhibition for CO + H_2 reaction can be accounted for by
two simultaneous explanations: (i) the active CO and H_2 species are
only those whose chemisorption sites are unaffected by SMSI (see
ref. (30) and close to the interface perimeter (31), and/or (ii)
SMSI does not exist any longer during the (CO + H_2) reaction, since
the catalyst returns to the "normal" state by partial reoxidation of
the support. In this last case, the exothermicity of titania's reoxidation can increase the catalyst temperature and accelerate the
initial reaction rate.

Photocatalytic cyclopentane–deuterium isotopic exchange (CDIE)

This reaction was chosen to illustrate the occurence of an electronic factor in a catalytic test under SMSI conditions. This reaction appears to be particularly suitable for this purpose since(i)
the photonic activation involves the formation of photoelectrons and
photoholes and (ii) the absence of oxygen-containing molecules as
well in the reactants as in the products allows to avoid the des-
truction of the SMSI state by a partial reoxidation of the support.

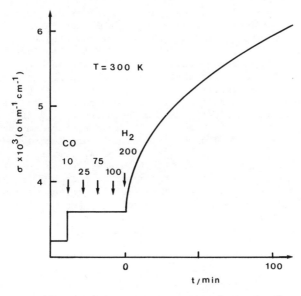

Figure 1. (CO + H$_2$) interactions on Pt/TiO$_2$ (5 %) in the SMSI state at room temperature (Pressures in Torr).

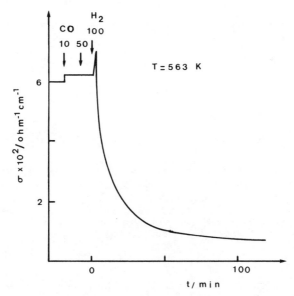

Figure 2. (CO + H$_2$) interactions on Pt/TiO$_2$ in the SMSI state at reaction temperature (290°C) (Pressures in Torr).

The reaction mechanism has been described in ref, (23), The reaction is carried out at temperatures $<-10°C$ where no dark thermal activity of the metal can be detected, It only occurs when near-UV light is admitted onto the solid, thus showing that the active phase is constituted by the support. However, the presence of a metal (Pt,Ni) is required to confer a catalytic character to the reaction since, otherwise, the reaction carried out on the bare support declines and stops after a certain time corresponding to an exhaustion of non-renewable active sites, These sites have been identified as deuterated hydroxyl groups since naked titania either dehydroxylated or pretreated in H_2 instead of D_2 is totally photo-inactive,

Moreover, from the electronic point of view, it has been shown by photoconductance measurements that (i) under vacuum the electrons photo-produced in the support are attracted by the metal because of the alignment of the Fermi levels corresponding to the illuminated state - the electron enrichment of the metal is supported by the fact that the higher the number of platinum particle, the higher the number of photo-electrons trapped and the smaller the photoconductance of the sample (33) - and (ii) under H_2, there still exists a reversible spill over of atomic hydrogen, described by a photoconductivity isotherm ($\sigma = a + b\, P_{H_2}^{1/2}$) formally identical to that obtained in the dark (Eq.7) but with "$\underline{a}$" being proportional to the radiant flux.

All these observations enabled us to propose a 8-step cyclic mechanism for the 100% selective formation of monodeurocyclopentane,

1°/ Dissociative chemisorption of D_2 on platinum
$$D_2 + 2\ Pt_s \rightleftharpoons 2\ Pt_s - D$$
and reversible weak chemisorption of cyclopentane on titania,
$$C_5 H_{10}\ (g) \rightleftharpoons C_5 H_{10}\ (ads)$$

2°/ Creation of electron-hole pairs by UV-light
$$(TiO_2) + h\nu \rightarrow e^- + p^+ \qquad (h\nu \geqslant E_g = 3\ eV)$$

3°/ Migration of photoelectrons to platinum
$$e^- + Pt \rightleftharpoons e^-_{Pt}$$

4°/ Reaction of photoholes with negatively charged OD^- groups
$$OD^-_s + p^+ \rightarrow OD^*_s$$

5°/ Deactivation of OD^* excited species by reaction with a cyclopentane molecule
$$OD^*_s + C_5 H_{10}\ (ads) \rightarrow O\overset{D}{\underset{H}{\cdots}}C_5 H_9 \rightarrow OH_s + C_5 H_9 D\ (g)$$

6°/ Reverse spill over of light hydrogen as proton from the support to the metal where it is reduced by excess electron (cathodic-like reduction).
$$H^+ + e^-_{Pt} \rightarrow H - Pt_s$$

7°/ Desorption of light hydrogen as HD and/or H_2 molecules into the gas phase.
$$Pt_S - H + Pt_s - D \rightarrow HD_{(g)} + 2\ Pt_S$$
$$2\ Pt_S H \rightarrow H_{2\,(g)} + 2\ Pt_s$$

8°/ Regeneration of photoactivable OD_s^- sites by direct spill over of deuterium atoms as deuterons at the surface of the oxide.

This cyclic reaction has been carried out more than 10^2 times on the same sites, under constant light flux, without any decrease in activity. This mechanism accounts for the 100% selectivity in C_5H_9D which constrasts with the polydeuteration found in classical catalysis on conventional Pt catalysts (32). It also accounts for the role of Pt which acts as a co-catalyst necessary (i) to separate the electron-hole pairs by attracting the electrons; (ii) to evolve light hydrogen in the gas phase and (iii) to regenerate the photo-activable sites.

Existence of an optimal metal content for CDIE due to an electronic factor

When platinum loading is varied on 2 orders of magnitude (0.1 ≪ Pt% ≪ 10%), but with a constant particle size ($\sim$ 2 nm), there exists an optimum content equal to $\sim$ 0.5 wt Pt% at which the rate of CDIE is maximum (Fig.3). This optimum is characteristic of the Pt/TiO₂ system since it has been found for other photocatalytic reactions, as well in the liquid or aqueous phase (dehydrogenation of $(C_1 - C_4)$ - alcohols (22) or of unsaturated ones (34)(curves (1) and (2) in Fig (3) as in the gas phase (photocatalytic isotopic exchange of oxygen (35)). Such an optimum is general and can be observed for other metals. It was found equal to $\sim$ 5 wt% for nickel (36) and seems to depend upon the nature and especially on the texture (particle size and dispersion) of the metal.

The ascending part of the curves in Fig.3 correponds to the beneficial role of the metal described in the above paragraph. The subsequent decrease in activity for higher metal contents cannot be accounted for by geometrical considerations (dark reverse reactions catalyzed by the metal, occultation of the photosensitive support by an increasing metallic phase...) since their influence is smaller than the effect observed. On the contrary the optimum metal content can be explained by an electronic effect since at high metal loading or high surface density of negatively charged crystallites, the positive photo-holes will progressively be preferentially attracted by these particles which become electron-hole recombination centers and hence inhibitors of the reaction.

CDIE in contact with Pt/TiO₂ catalysts under SMSI conditions

When CDIE is repeated on the same Pt/TiO₂ catalysts but reduced at 500°C, one observes a strong decrease in activity (curve B, Fig.3). This might be partially explained by a reduction of deuteroxyl group coverage. However, the lacking OD⁻ groups would be readily replaced by deuterium spillover provided its chemisorption on platinum be not totally inhibited by SMSI. Moreover, the decrease in OD⁻ coverage would have affected the photocatalytic activity in CDIE independently of the Pt percentage. In fact, the smaller the platinum loading-with a constant particle size -, the stronger the effect of SMSI quantified by the "inhibition factor R", defined as the ratio of the "normal" reaction rate, i.e. for T_R < 300°C, to the "SMSI" reaction rate (Fig.4). For very low Pt% ($\sim$ 0.1%), R could not be determined accurately since r_{SMSI} was too close to that obtained on naked ti-

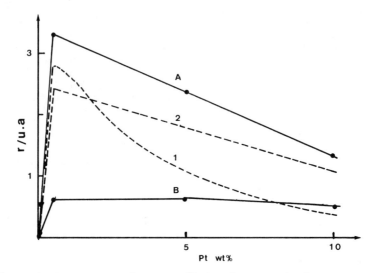

Figure 3. Photocatalytic rates (in arbitrary units) vs Pt loading. Curves (A) and (B) : CDIE in "normal" and "SMSI" state ; dashed curves (1) and (2) : 1 - propanol and allyl-alcohol dehydrogenation.

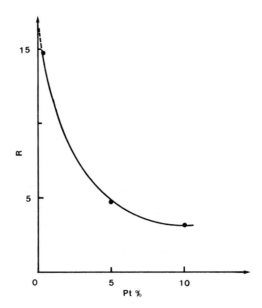

Figure 4.   Inhibition factor R vs Pt%.
R = rate (normal state)/rate (SMSI).

tania in the "normal" state. This result clearly shows that the extent of SMSI is relevant to the quantity of metal deposited, as observed for hydrogen chemisorption (Table I). If the migration of TiO_x species on the metal particles were the major factor responsible for the decrease in activity, it would not be metal content sensitive, as observed in Fig.4, since the support particle represent a large enough reservoir of oxide or suboxide species to cover or decorate all the metal crystallites. On the contrary, it was shown for Pt and Ni that the higher the metal content, the smaller the free (photo-) electron density in the support (33,36). Consequently a catalyst having a low Pt content would be more perturbed electronically and thence would move more easily to the SMSI state than one with a higher Pt content.

Conclusions

The SMSI state of M/TiO_2 catalysts have been examined from the electronic point of view in three different situations: (i) chemisorptive capacity, (ii) $(CO + H_2)$ interactions and (iii) CDIE photocatalytic test. In all cases, the free electron density was involved. Without rejecting any influence of the decoration of the metal particles by suboxide species, the electronic effect can be considered as responsible – at least partially – for the SMSI state (11,21).

The excess electron in the metal could not be quantified from electrical conductivity data. On electrostatic grounds, it does not seem possible to transfer a great number of electrons sufficient to reach unity for e/M. This is confirmed by simple considerations on the reduction level of titania and the relative molar ratio n_{Pt}/n_{TiO_2} . It seems more reasonable to think in terms of "long" distance influence. For instance, since H_2 and CO chemisorb on the metal surface as dipoles with a donor character, even a small excess of electrons in the metal or at its interface with the support will counteract this chemisorption.

For CDIE, besides the inhibition of deuterium chemisorption necessary to regenerate OD^- photoactivable sites, the electron excess in the metal increases the recombination of holes which is detrimental for the basic activation step of the reaction.

The electronic metal-support interactions can also account for the lesser extent of SMSI with respect to H_2 chemisorption and catalysis when the metal loading is increased at constant crystallite size, since for an equivalent quantity of free electrons created by reduction in the support, the higher number of metal atoms present leads to weaker electronic perturbations per atom concerned.

The electron transfer from a n-type semiconductor support to a metal is quite general and can be observed in milder conditions when high temperature reduction is replaced by suitable illumination at room temperature. It has been detected either under vacuum or in hydrogen by photoconductance measurements (33) and even in oxygen by photosorption determinations (35).

More work on SMSI has to be done to elucidate the exact respective part - if any - of decoration and electron effects and a possible technique for that would be heat flow microcalorimetry.

Acknowledgments: The author thanks Dr. Pichat for his interest, and MM. J. Disdier and H. Courbon for technical assistance (photoconductance and CDIE analyses respectively).

Literature cited

1. Tauster, S.J.; Fung, S.C.; Garten, R.L. J. Amer. Chem. Soc.
    1978, 100, 170.
2. Tauster S.J.; Fung S.C.; Baker, R.T.K.; Horsley, J.A. Science
    1981, 211, 1121.
3. "Metal–Support and Metal–Additive Effects in Catalysis", Imelik,
    B. et al. Eds, Elsevier : Amsterdam, 1982.
4. "Symposium on Strong–Metal–Support Interactions"; American Che-
    mical Society : Washington D.C., 1985.
5. Horsley, J.A.; J. Amer. Chem. Soc.1979, 101, 2870.
6. Bahl, M.K.; Tsai, S.C.; Chung, Y.W. Phys. Rev. 1980, B21, 1344.
7. Kao, C.C.; Tsai, S.C.; Bahl, M.K.; Chung, Y.W. Surf. Sci. 1980,
    95, 1.
8. Sanchez, J.; Koudelka, M.; Augustinki, J.; J. Electroanal. Chem.
    1982, 140, 161.
9. Kao, C.C.; Tsai, S.C.; Chung, Y.W. J. Catal. 1982, 73, 136.
10. Bor Her Chen; White, J.M. J. Phys. Chem. 1983, 81, 147.
11. Herrmann, J.M. J. Catal. 1984, 89, 404.
12. Conesa, J.C.; Malet, P.; Muñoz, A.; Munuera, G.; Sainz, M.T.;
    Sanz, J.; Soria, J. Proc. 8th Int. Cong. Catal.; Verlag Chemie:
    Weinheim, 1984, Vol.5, 217.
13. Li Wenzhao; Chen Yixuan; Yu Chunying; Wang Xiangzhen; Hong Zupei;
    Wei Zhobin; ibid. p. 205.
14. Mériaudeau, P.; Dutel, J.F.; Dufaux, M.; Naccache, C. "Metal-
    Support and Metal–Additive Effects in Catalysis", Imelik, B. et
    al. Eds, Elsevier: Amsterdam, 1982, p. 95.
15. Santos, J.; Phillips, J.; Dumesic, J.S. J.Catal.1983, 81, 147.
16. Resasco, D.E.; Haller G.L. J. Catal. 1983, 82, 279.
17. Sadeghi, H.R.; Henrich, V.E. J. Catal. 1984, 87, 279.
18. Baker, R.T.K.; Prestridge, E.B.; Mc. Vicker, G.B. J. Catal.1984,
    89, 422.
19. Haller, G.L.; Henrich, V.E.; Mc Millan, M.; Sadeghi, H.R.;
    Sakellan, S.; Proc. 8th Int. Cong. Catal., Verlag Chemie:
    Weinheim, 1984, Vol. 5, 135.
20. Demmin, R.A.; Ko, C.S.; Gorte, R.J., J. Phys. Chem. 1985, 89,
    1151.
21. Belton, D.N.; Sun Y.-M.; White, J.M. J. Phys. Chem. 1984, 88,
    5172.
22. Pichat, P.; Mozzanega, M.-N., Disdier, J.; Herrmann, J.-M. Nouv.
    J. Chim. 1982, 6, 559.
23. Courbon, H.; Herrmann, J.-M.; Pichat, P. J.Catal. 1981, 72, 129.
24. Herrmann, J.-M.; Pichat, P.; J. Catal. 1982, 78, 425.
25. Baker, R.T.K. "Metal–Support and Metal–Additive Effects in Ca-
    talysis", Imelik, B. et al. Eds, Elsevier: Amsterdam, 1982,p.37.
26. Yamamoto, N.; Tonomura, S.; Matsuoka, R.; Tsubomura, H. J. Appl.
    Phys.1981, 52, 6227.
27. Aspnes, D.E.; Heller, A. J. Phys. Chem. 1983, 87, 4919.
28. Gerischer, H. J. Phys. Chem. 1984, 88, 6096.
29. Vannice, M.A.; Moon, S.H.; Twu, C.C. "Surface Studies Related to
    Petroleum Chemistry"; American Chemical Society : Houston, 1980,
    p.303.
30. Vannice, M.A.; Twu, C.C.; Moon, S.H. J. Catal. 1983, 79, 70.
31. Burch, R.; Flambard, A.R. J. Catal. 1982, 78, 389.

32. Inoue, Y.; Herrmann, J.-M.; Schmidt, H.; Burwell, R.L.; Butt,
    J.B.; Cohen, J.B. J. Catal. 1978, 53, 401.
33. Disdier, J.; Herrmann, J.-M.; Pichat, P.; J. Chem. Soc. Faraday
    Trans. 1 1983, 79, 651.
34. Pichat P.; Disdier, J.; Mozzanega, M.-N.; Herrmann, J.-M.; Proc.
    8th Int. Cong. Catal., Verlag Chemie: Weinheim, 1984, Vol. 3,487.
35. Courbon, H.; Herrmann, J.-M.; Pichat, P. J. Phys. Chem.1984, 88,
    5210.
36. Prahov, L.T.; Disdier,J.; Herrmann, J.-M.; Pichat P.; Int. J.
    Hydrogen Energy 1984, 9, 397.

RECEIVED September 12, 1985

21

Spectroscopic and Electrochemical Study of the State of Pt in Pt–TiO₂ Catalysts

M. Spichiger-Ulmann, A. Monnier, M. Koudelka, and J. Augustynski

Département de Chimie Minérale, Analytique et Appliquée, Université de Genève, 1211 Genève 4, Switzerland

Previous studies performed in this laboratory have led to the observation of a strong metal-support interaction (SMSI) for a Pt/TiO₂ catalyst heated in argon (i.e., in the absence of hydrogen) at temperatures higher than ∿ 500°C. This SMSI effect, evidenced by a negative shift of binding energy for the Pt4f electrons (reaching 0.6 eV), presents the peculiarity of being persistent even after exposure of the samples to air for several days. Such a behavior, observed for the titanium supported Pt/TiO₂ films, is thought to arise from the fact that Ti atoms from the underlying metal may act as a reducing agent during the thermal treatment. Because of a relatively small specific surface area of this Pt/TiO₂ catalyst, an electrochemical technique - the cyclic voltammetry - was chosen to characterize the properties of platinum on the surface of various SMSI and non-SMSI samples. Results regarding H and CO chemisorption, obtained for different Pt/TiO₂ films reduced or oxidized at various temperatures, are discussed. A tentative explanation of the SMSI effect exibited by the Pt/TiO₂ films is proposed.

A large number of studies, stirred up with the report in 1978 by Tauster et al. (1) of strong metal-support interaction (SMSI) occurring for a series of TiO₂-supported noble metal catalysts, has allowed successive refinements of the model of such a system. Both, geometrical and electronic factors are presently considered between possible causes of the altered adsorption behavior of SMSI catalysts, appearing in a quasi-total suppression of hydrogen and carbon monoxide chemisorption. Following an earlier suggestion of Meriaudeau et al. (2), the high-temperature reduction of the catalyst, inducing an SMSI behavior, has been conclusively shown (3,4) to lead to at least partial recovering of the metal particles by the reduced oxide of the support. However, it is still not quite clear whether the encapsulation of metal particles is the cause or the consequence of an electronic interaction with the support.

0097–6156/86/0298–0212$06.00/0
© 1986 American Chemical Society

The existence of the electronic interaction in a model Pt/TiO$_2$ system has been inferred (5,6) from our observation of a significant and reproducible negative binding energy (BE) shift for core level Pt electrons, following the reduction of the catalyst in argon at 550°C. The kind of samples used in these studies (5,6), consisting of Pt/TiO$_2$ films supported on Ti metal, allowed to minimize charging effects which usually render difficult precise interpretation of the photoelectron spectra for powder catalysts. Apart from the extent of the observed Pt4f BE shift, which reached -0.6 eV in comparison with the BE value measured for a Pt foil, it is its persistence, in spite of exposure of the samples to air, which constitutes the most peculiar feature of these Pt/TiO$_2$ films. In this paper we discuss in more details the morphological characteristics and adsorption properties of this SMSI Pt/TiO$_2$ system in comparison with those of other non-SMSI Pt/TiO$_2$ films. In particular, the second part of this paper focuses on the informations, regarding chemisorption of hydrogen and of carbon monoxide, available from cyclic-voltammetric measurements at Pt/TiO$_2$ surfaces.

Experimental

Different kinds of platinum catalysts investigated in this study were supported on 10 to 15 μm thick TiO$_2$ films. These films were prepared according to a standard procedure (6,7) involving a layer by layer hydrolytic decomposition of an alcoholic TiCl$_4$ solution, applied onto metallic titanium substrates. After each application the specimens were heated at 450°C in air for 15 min.; following the deposition of the last layer the heating was prolonged up to 1 h. Thermal platinum deposits, referred as Pt(th)/TiO$_2$, were obtained by impregnating the TiO$_2$ films with an aqueous solution of H$_2$PtCl$_6$ and decomposing it in air, in most cases at 450°C. The platinized TiO$_2$ films were then subjected for 40 min. to annealing in ultra-pure argon, in general at 300° or 550°C. The latter high-temperature reduction treatment led to an SMSI type behavior.

The TiO$_2$ films which served as supports for photochemical (6,8) and electrochemical (9) platinum deposits were, prior to recovering them with Pt, reduced in argon under conditions identical with those used for SMSI Pt(th)/TiO$_2$ samples, i.e., at 550°C and during 40 min.

X-ray photoelectron spectroscopic (XPS) measurements were taken on a Varian IEE-15 spectrometer, using MgKα$_{1,2}$ radiation at 1253.6 eV. Basic pressure in the sample chamber of this spectrometer is not better than 10^{-7} torr. All XPS analyses were performed ex situ, the samples being transferred to the spectrometer through air. Binding energies were referenced to a main C1s line, due to residual pump oil on the sample surface, taken at 285 eV. The Au4f(7/2) line of metallic gold served as a second internal reference. The BE difference between these two levels was remarkably constant, 201.2 ± 0.1 eV, in agreement with the results of an extensive comparative study published in the literature (5). Relative surface concentrations of different species were determined as described previously (10).

Several portions of the powder catalyst, detached from the surface region of Pt(th)TiO$_2$ specimens, were examined by transmission

electron microscopy (TEM). Both, samples ultrasonically dispersed in
alcohol and extractive replica with isolated platinum particles were
used.

Titanium-supported Pt/TiO_2 specimens characterized by cyclic
voltammetry had an apparent (geometrical) area of 0.28 cm^2. The
measurements were carried out in an aq. 0.5 M $NaHSO_4$/0.5 M Na_2SO_4
solution kept at 20°C. The solution was saturated with Ar or CO under
1 atm. The potentials were monitored with respect to and are referred
to the reversible hydrogen electrode (RHE) in the same solution. The
cyclic-voltammetry procedure followed that conventionally employed
with the potential swept, in general, at a rate of 0.1 $V \cdot s^{-1}$.

Results and discussion

Structural studies

Both, X-ray diffraction and electron micro-diffraction analyses
showed the outer part of the reduced TiO_2 films to consist mainly
of polycrystalline anatase domains with only small adjunction of
rutile. No structural changes were visible for high-temperature-
reduced $Pt(th)/TiO_2$ films for which, besides anatase, metallic platinum was also observed. Due to a particular mode of reduction of
TiO_2 and $Pt(th)/TiO_2$ samples used in this study, the interior of the
films, close to the interface with titanium metal, is expected to
consist of lower titanium oxides including well-conducting TiO and
Ti_2O_3. This was qualitatively confirmed by the presence in this
region of the film of perceptible oxygen and titanium concentration
gradients detected by means of electron microprobe (6). However,
limited resolution of the latter method does not allow detailed interpretation of the observed concentration profiles. The reduction
at 550°C of the titanium oxide layer is supposed to proceed mainly
from the metal-oxide interface and to involve both the Ti migration
into TiO_{2-x} and the O^{2-} migration into the metal (Figure 1). As a
result, two kinds of defects, the oxygen vacancies, V_O, and the
titanium atoms in interstitial positions, $(Ti)_i$, may be expected to
coexist (11) in thus reduced TiO_2 films.

An extensive XPS study, involving large number of TiO_2 samples
platinized using thermal, photochemical and electrochemical methods,
showed that, characteristically, Pt4f binding energies corresponding
to the $Pt(th)/TiO_2$ samples reduced at 550°C were substantially lower
than those for bulk platinum metal and those for other Pt/TiO_2
samples. The main results of these measurements are summarized in
Table I.

In order to avoid the uncertainties associated with a possible
contribution to measured BE's from the extra-atomic relaxation
energy (6,12) we chose for this comparison a relatively thick photochemical Pt deposit, $Pt(ph)/TiO_2$ (in fact, the Ti signal from the
TiO_2 support could not be distinguished in that case). The corresponding Pt4f(7/2) BE was of 71.4 eV, i.e. perceptibly higher than that
of unsupported bulk Pt. As a rule, Pt4f(7/2) BE's slightly higher
than 71.1 eV were also found for other $Pt(ph)/TiO_2$ samples containing
less or much less platinum, like the one shown in Figure 2. Platinum
appears on the low-magnification scanning electron micrograph as
irregular islands dispersed on the TiO_2 surface and in the cracks.

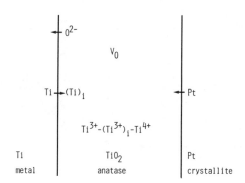

Figure 1. Diffusion processes expected to occur in the Ti-suppor-
ted, platinized TiO_2 film during annealing in argon at 550°C.

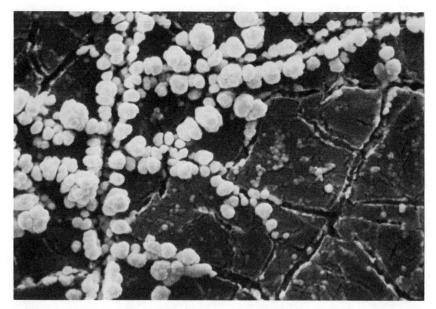

Figure 2. Scanning electron micrograph (magnification x 5000)
of photochemically platinized, prereduced TiO_2 film.

Table I. Surface composition and XPS binding energies for TiO_2 and Pt/TiO_2 films, and for bulk Pt.

Sample	BE, eV			surface atomic ratio	
	Pt4f(7/2)	Ti2p(3/2)	O1s*	Pt/Ti	O/Ti
TiO_2(Ar, 550°C)		458.8	530.2		2.05
Pt(ph)/TiO_2	71.4 (δ=1.4)	--	532.1		
Pt(th)/TiO_2(Ar, 550°C)	70.5 (δ=1.4)	458.7	530	0.1	2.03
smooth Pt foil	71.1 (δ=1.5)		532.3		

*The O1s BE's and O/Ti ratios are given for the most intense oxygen 1s signal.

Table II. Surface composition and XPS binding energies for various Pt(th)/TiO_2 (Ar, 550°C) samples.

Sample No	BE, ev			surface atomic ratio	
	Pt4f(7/2)	Ti2p(3/2)	O1s	Pt/Ti	O/Ti
1	70.5 (δ=1.6)	458.8	530	0.015	1.96
2	70.5 (δ=1.5)	458.8	530	0.04	2.04
3	70.5 (δ=1.4)	458.7	530	0.1	2.03
4	70.6 (δ=1.6)	458.8	530	0.04	2
5	70.6 (δ=1.5)	458.8	530	0.07	1.92
6	70.6 (δ=1.5)	458.8	530	0.15	1.95
7	70.7 (δ=1.5)	458.9	530	0.13	1.93
8	70.9 (δ=1.6)	458.9	530.1	0.13	1.75
9	70.9 (δ=1.5)	458.9	530.1	0.23	2.04

If one assumes that platinum, present on the surface of Pt(th)/TiO$_2$ samples during their annealing in argon, did not affect in a significant way the extent of reduction of the TiO$_2$ support, the Pt4f BE difference occurring between Pt(ph)/TiO$_2$ and Pt(th)/TiO$_2$ samples (equal to 0.9 eV) can be assigned neither to relaxation energy nor, more generally, to a matrix effect (6,13). On the contrary, since the size of Pt particles observed with TEM for high-temperature-reduced (HTR) Pt(th)/TiO$_2$ deposits (Figure 3) was much smaller than that of the photochemically formed Pt islands shown in Figure 2, the actual chemical BE shift could possibly be even larger than the experimentally observed Pt4f BE shift (6,12). However, it is to be noted that despite a really heterogeneous metal particle size distribution characterizing the above mentioned Pt(th)/TiO$_2$ deposits, the corresponding Pt4f photoelectron signals had the same width at half-maximum (indicated in Tables as δ) as the signals recorded for Pt foil (Tables I and II). This is inconsistent with diversified contributions to the Pt4f electron spectra and thus renders doubtful perceptible differences in relaxation energy between the Pt/TiO$_2$ samples examined during this study.

XPS measurements performed with a large number of HTR Pt(th)/TiO$_2$ samples, containing different amounts of platinum showed certain dispersion of the Pt4f(7/2) BE's examplified in Table II. The values ranged from 70.5 to 70.9 eV (respectively 0.6 to 0.2 eV lower than for Pt foil) and tended to increase for the Pt/Ti ratios higher than 0.1-0.15. In contrast with distinct differences between the Pt4f BE's for the HTR thermal and the photochemical Pt/TiO$_2$ deposits, the Ti2p(3/2) BE's listed in Tables I and II (including that for non-platinized TiO$_2$) all range from 458.7 to 458.9 eV. Also the O1s BE (corresponding to the main signal due to O^{2-} ions in the TiO$_2$ lattice) was almost constant at 530-530.1 eV. Importantly, there was no significant change in the positions of Ti2p and O1s signals consecutive to HTR treatment of the Pt/TiO$_2$ film, following which the Pt4f(7/2) line shifted 0.4 eV to lower BE's (Table III). Table III examplifies also an unexpected and interesting effect exhibited by a few samples, namely the persistence of the low BE Pt4f(7/2) value in spite of prolonged (in one case, for 75 days) exposure to air. Still, another Pt(th)/TiO$_2$ specimen, maintained in air for 4 months, showed finally a positive shift of the Pt4f(7/2) BE (Table IV). This shift was accompanied by a significant increase, up to 2 eV, of the width at half-maximum (WHM) of the corresponding photoelectron signal (Figure 4). Since, as just mentioned, the WHM of the Pt4f(7/2) lines were very reproducible, close to 1.5 eV, a value of 2 eV can be taken as an evidence for the presence of non-equivalent platinum atoms on the surface of that partly "oxidized" sample. Other HTR Pt(th)TiO$_2$ samples exposed to air for a similar period of time exhibited a shoulder to the main Pt4f signal, due to Pt(II) oxide (in comparison, the oxidation in air of the non-SMSI samples was in general much more rapid, needing few days).

As shown in Table IV, the initial, lower Pt4f BE could be restored by high-temperature reduction in argon and then increased

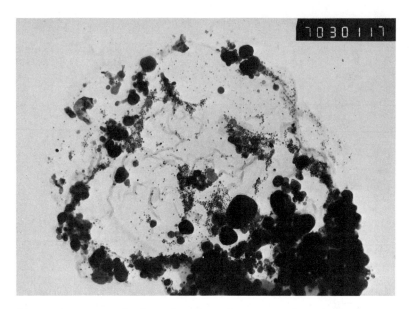

Figure 3.   Transmission electron micrograph (replica) of Pt
particles isolated from the Pt(th)/TiO$_2$ film annealed in
argon at 550°C.

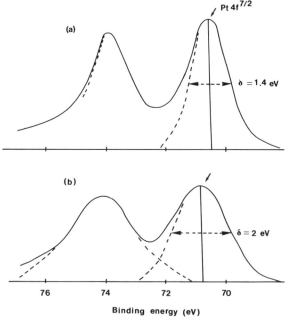

Figure 4.   X-ray photoelectron spectra of the Pt4f level for the
reduced (550°C) Pt(th)/TiO$_2$ deposit : (a) as prepared, (b) after
4 months of exposure to air.

Table III.   Surface composition and XPS binding energies

Sample	BE, eV			surface atomic ratio
	Pt4f(7/2)	Ti2p(3/2)	O1s	Pt/Ti
a) Pt(th)/TiO$_2$(air, 450°C) before annealing in Ar	71 (δ=1.5)	458.8	530.1	0.04
b) Pt(th)/TiO$_2$ (Ar,550°C) sample (a) after reduction	70.6 (δ=1.6)	458.8	530	0.04
c) Pt(th)/TiO$_2$ (Ar,550°C) sample (b) 75 days later	70.6 (δ=1.5)	458.7	529.9	0.03
d) Pt(th)/TiO$_2$ (Ar,550°C) sample (c) after 2nd prolonged reduction in Ar	70.6 (δ=1.5)	458.7	530	0.03

Table IV.   Surface composition and XPS binding energies

Sample Pt(th)/TiO$_2$(Ar,550°C)	BE, eV			surface atomic ratio
	Pt4f(7/2)	Ti2p(3/2)	O1s	Pt/Ti
a) as prepared	70.5 (δ=1.4)	458.7	530	0.1
b) after a 4 months exposure to air	70.9 (δ=2)	458.9	530.1	0.1
c) after annealing in Ar at 550°C	70.6 (δ=1.5)	458.7	530	0.09
d) after annealing in air at 450°C	70.9 (δ=1.7)	458.8	530	0.09

again by oxidation in air at 450°C. The same partial reversibility
of BE changes was also observed consecutive to electrochemical
oxidation and reduction of the Pt(th)/TiO$_2$ samples in an aq.sodium
sulfate solution at room temperature (6).

Electrochemical investigations

The usefulness of cyclic voltammetry as a tool for characterization
of noble metal catalysts has been demonstrated by several authors
(14–17). The application of this method requires the metal/support
system to be electrically conducting, the condition which has been
shown to be fulfilled by the reduced Pt/TiO$_2$/Ti samples (5).
      When the potential of a smooth platinum electrode immersed in
an acidic aq. solution is swept in cathodic direction, two reduction
peaks, due to formation of atomic hydrogen, are usually observed at
potentials preceding H$_2$ evolution (Figure 5). Three oxidation peaks
are visible on the reverse(anodic) sweep, corresponding to removal
of H atoms from different submonolayer adsorption states. It is to
be mentioned that four states of H adsorption at Pt could be resolved
on cyclic voltammograms recorded in dilute acid solutions, in the
absence of significant specific adsorption of anion (18). The occurrence of these multiple states of H chemisorption has been explained
(18) by a combination of intrinsic and induced heterogeneity factors.
The former are mainly associated with the presence of various crystal
planes on the polycrystalline Pt surface whilst the latter arise
from progressive increase of H coverage.
      The region of the cyclic voltammogram, corresponding to anodic
removal of H$_{ad}$ atoms, looks quite similar to the thermal desorption
spectra of platinum catalysts. However, unlikely the thermal desorption spectra, the cyclic-voltammetric profiles for H chemisorbed on
Pt are usually free of kinetic effects. In addition, the electrochemical techniques offer the possibility of cleaning eventual impurities from the platinum surface through a combined anodic oxidation-
cathodic reduction pretreatment. Comparative gas-phase and electrochemical measurements, performed for dispersed platinum catalysts,
have previously demonstrated similar hydrogen and carbon monoxide
chemisorption stoichiometries at both the liquid and gas-phase
interfaces (14).
      Cyclic voltammograms recorded for a series of photochemically
and electrochemically platinized TiO$_2$ samples (Figure 6) exhibited
principal features typical of smooth platinum. There were some minor
differences, especially for small Pt coverages, regarding the distribution of H atoms amongst various adsorption states or the position
of the Pt oxide reduction peak. In contrast, the behavior of HTR
Pt(th)/TiO$_2$ electrodes was diametrically different, the multiple H
adsorption peaks being replaced by smooth increase of the current
both in the cathodic and anodic directions (Figure 7). The voltammograms of this kind are known, for example, for electrochromic WO$_3$
films (19) and are associated with the proton injection into, respectively, the proton removal from the oxide lattice. Prereduced TiO$_2$
films also display similar behavior but the corresponding cathodic

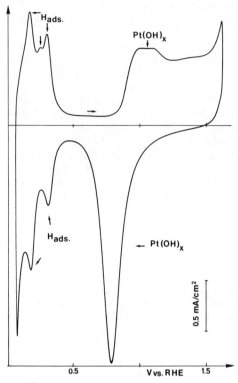

Figure 5. Typical cyclic voltammogram of smooth platinum obtained in a deaerated acidic sulfate solution.

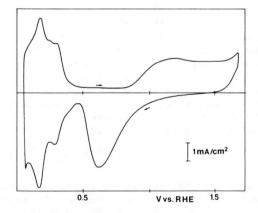

Figure 6. Cyclic voltammogram of a Pt(ph)/TiO$_2$ film electrode, recorded in a 0.5 M NaHSO$_4$/0.5 M Na$_2$SO$_4$ solution saturated with argon.

and anodic peaks are shifted to much more negative, non-equilibrium potentials (Figure 7, dotted line curve). This suggests that, in the case of HTR Pt(th)/TiO_2 electrodes, the role of Pt crystallites is to mediate entering of protons into the TiO_2 film and/or to enhance the film conductivity. Voltammetric experiments, involving polarization of the HTR Pt(th)/TiO_2 electrode at constant potential corresponding to the end of the cathodic sweep, before starting the anodic sweep, showed that the proton injection was restricted only to the surface of the film. The anodic oxidation peak recorded in the latter experiment was, in fact, unchanged with respect to that observed during continuous potential cycling. These features of electrochemical behavior of the HTR Pt(th)/TiO_2 samples are to be associated with the results of gas-phase experiments indicating that, whilst SMSI Pt catalysts become inactive with respect to H_2 dissociation, they are still able to catalyse reduction of the TiO_2 support by H atoms (C. Naccache, this Symposium). It should also be mentioned that a Pt(th)/TiO_2 electrode, which had been subjected to the annealing in argon at 300°C, i.e., below the critical temperature necessary for inducing SMSI properties, showed voltammetric profiles (Figure 8, dotted line curve) of the kind of those exhibited by the Pt(ph)/TiO_2 and smooth Pt electrodes.

The differences in behavior of the HTR Pt(th)/TiO_2 SMSI samples and of other non-SMSI Pt/TiO_2 films were even more marked in the case of carbon monoxide chemisorption experiments. In Figure 9 is shown a typical cyclic voltammogram recorded for a smooth Pt electrode in the solution saturated with CO. The strong adsorption of CO results both in a quasi-total suppression of the current in the hydrogen region and in the appearance of a sharp anodic peak, at ~ 0.9 V vs.RHE, corresponding to the removal of CO_{ad} species from the Pt surface. The latter peak, associated with a two-electron oxidation of adsorbed carbon monoxide, usually serves as a basis for estimating the coverage of the electrode by CO_{ad}. Again, the behavior of Pt(ph)/TiO_2 and of LTR (Ar,300°C) Pt(th)/TiO_2 samples (Figures 10 and 11, respectively) was similar to that of smooth Pt. On the other hand, in the case of HTR Pt(th)/TiO_2 electrodes, the anodic peak corresponding to the CO_{ad} oxidation was absent from the voltammogram (Figure 12). In addition, the latter voltammogram, in contrast with those for the non-SMSI Pt/TiO_2 samples, shows pronounced cathodic and anodic currents in the hydrogen region, which are also inconsistent with strong CO adsorption on the surface of HTR Pt(th)/TiO_2 samples. These currents were, however, less intense than those observed in the absence of CO, suggesting the occurrence of an apparently weak adsorption of carbon monoxide.

Conclusions

First and the most obvious conclusion, which can be drawn from the experiments described above, is that the SMSI state of Pt and other noble metals dispersed on a TiO_2 support can be induced, probably more effectively, by means of the reduction with Ti metal. Consequently, the hydrogen spillover is not the only mechanism leading to SMSI.

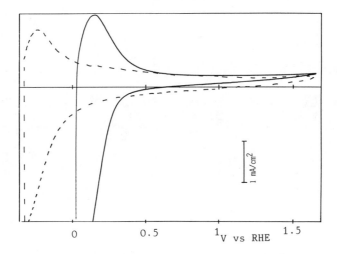

Figure 7. Cyclic voltammograms for Pt(th)/TiO$_2$ (solid line curve) and TiO$_2$ (dotted line curve) films reduced in argon at 550°C. Conditions identical as in Figure 6.

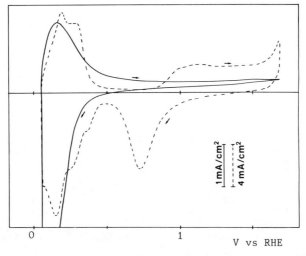

Figure 8. Comparison of cyclic voltammograms for Pt(th)/TiO$_2$ films reduced, respectively, at 300°C (dotted line curve) and at 550°C (solid line curve). Conditions as in Figure 6.

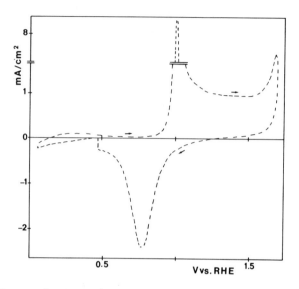

Figure 9. Typical cyclic voltammogram of smooth Pt recorded in
a 0.5 M NaHSO$_4$/0.5 M Na$_2$SO$_4$ solution saturated with CO.

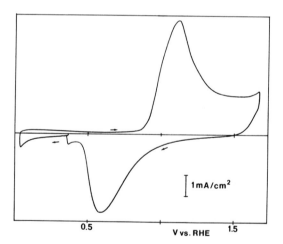

Figure 10. Cyclic voltammetric current–potential profile for
a Pt(ph)/TiO$_2$ film electrode in the presence of CO; conditions
as in Figure 9.

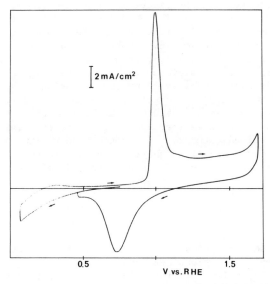

Figure 11. Cyclic voltammogram for a Pt(th)/TiO$_2$ film reduced at 300°C, recorded in the presence of CO. Conditions like in Figure 9.

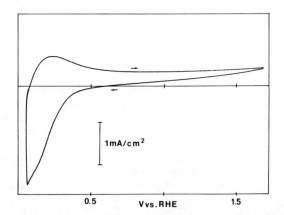

Figure 12. Cyclic voltammogram for a HTR Pt(th)/TiO$_2$ film in the solution containing CO. Conditions as in Figure 9.

One possible explanation of the results of electrochemical measurements obtained with the HTR Pt(th)/TiO$_2$ samples may involve encapsulation of the metal particles in the TiO$_2$ support or their partial recovering by TiO$_{2-x}$. This would be consistent with the Pt migration, in-depth the film, revealed by electron microprobe analyses of a film cross-section (6). However, it is not quite clear if such a migration does not mainly proceed through the cracks in the film surface, which are visible on the SEM image (Figure 2). On the other hand, diffusion processes occurring in the surface region of the Pt/TiO$_2$ films during the reduction treatment can hardly be responsible for the large and persistent BE shift of Pt4f photoelectrons. The latter, as suggested earlier (20,21), can be explained in terms of the Fermi level adjustment, between the reduced TiO$_2$ and Pt, leading to the formation of a Schottky barrier. In this connection, it is worthy to mention recent measurements (22) of the work function changes resulting from the adsorption of Pt on the 001 face of reduced rutile. An increase of the work function by up to 1 eV, corresponding to a Pt coverage equivalent to 2 monolayers, has been observed. The change of the work function for a 1 monolayer Pt coverage was of approximately 0.7 eV.

However, the assumption of the Schottky barrier formation raises some complex questions regarding both the size of particles on the surface of which (in principle, in a region of less than 0.5 nm) a significant change of electron density may be expected, and the nature of the space charge layer on the oxide side of the barrier. The optimum conditions would arise for the size of Pt particles comparable with the thickness of the space charge layer. Such an adjustment would certainly be favored by the Pt encapsulation and by a decrease of the space charge layer thickness, around Pt particles, due to the increase of the donor concentration.

There is yet an additional feature which could explain both the long-term persistence of the negative Pt4f BE shift and the mechanism by which the electron transfer to the surface of Pt particles may take place. Because of the pronounced tendency of titanium dioxide to dissociate water, the exposure of prereduced Pt/TiO$_2$ to ambient atmosphere is expected to lead to the spontaneous formation of TiOOH, through the proton injection into the surface region of the sample. The protons can be easily accommodated in vacant octahedral sites in the anatase lattice, compensating the excess charge due to the presence of Ti^{3+} ions (d^1). Assuming that these hydrogen ions (with an effective fractional positive charge) are present around Pt particles encapsulated close to the catalyst surface, they may both stabilize negative excess charge in platinum and mediate the 3d^1 electron transfer from the Ti^{3+} ions to the Pt crystallites. Such a "triple layer" model would be consistent both with our results of XPS measurements for the reduced Pt(th)/TiO$_2$ samples pre-exposed to air and with the observation of the negative Pt4f BE shift consecutively to the electrochemical reduction of such a sample (5).

Acknowledgments

We are grateful to Mrs. C. Leclercq (Institut de Recherches sur la Catalyse, Villeurbanne) for obtaining transmission electron micrographs.

Literature Cited

1. Tauster, S.J.; Fung, S.C.; Garten, R.L. J. Am. Chem. Soc. 1978, 100, 170.
2. Meriaudeau, P.; Dutel, J.F.; Dufaux, M.; Naccache, C. In "Metal-Support and Metal-Additive Effects in Catalysis"; Imelik, B.; Naccache, C.; Coudurier, G.; Praliaud, H.; Meriaudeau, P.; Gallezot, P.; Martin, G.A.; Vedrine, J.C. Eds.; STUDIES IN SURFACE SCIENCE AND CATALYSIS SERIES No 11, Elsevier : Amsterdam, 1982; p. 95.
3. Jiang, X-Z.; Hayden, T.F.; Dumesic, J.A. J. Catal. 1983, 83, 168.
4. Simoens, A.J.; Baker, R.T.K.; Dwyer, D.J.; Lund, C.R.F.; Madon, R.J. J. Catal. 1984, 86, 359.
5. Sanchez, J.; Koudelka, M.; Augustynski, J. J. Electroanal. Chem. 1982, 140, 161 and references therein.
6. Koudelka, M.; Monnier, A.; Sanchez, J.; Augustynski, J. J. Mol. Catal. 1984, 25, 295.
7. Stalder, C.; Augustynski, J. J. Electrochem. Soc. 1979, 126, 2007.
8. Lehn, J.-M.; Sauvage, J.-P.; Ziessel, R. Nouv. J. Chim. 1980, 4, 623.
9. Khazova, O.A.; Vasiliev, Y.B.; Bagotskii, V.S. Elektrokhimiya 1970, 6, 1367.
10. Koudelka, M.; Sanchez, J.; Augustynski, J. J. Electrochem. Soc. 1982, 129, 1186.
11. Goodenough, J.B. "Les Oxydes des Métaux de Transition"; Gauthier-Villars : Paris, 1973; p. 282.
12. Takasu, Y.; Unwin, R.; Tesche, B.; Bradshaw, A.M.; Grunze, M. Surf. Sci. 1978, 77, 219.
13. Kim, K.S.; Winograd, N. Chem. Phys. Lett. 1975, 30, 91.
14. Bett, J.; Kinoshita, K.; Routsis, K.; Stonehart, P. J. Catal. 1973, 29, 160.
15. Kinoshita, K.; Lundquist, J.; Stonehart, P. J. Catal. 1973, 31, 325.
16. Scortichini, C.L.; Reilley, C.N. J. Catal. 1983, 79, 138.
17. Mahmood, T.; Williams, J.O.; Miles, R.; McNicol, B.D. J. Catal. 1981, 72, 218.
18. Angerstein-Kozlowska, H.; Sharp, W.B.A.; Conway, B.E. Proc. Symp. on Electrocatalysis; Breiter, M.W., Ed.; The Electrochem. Soc. : Princeton, N.J., 1974; p. 94.
19. Reichman, B.; Bard, A.J. J. Electrochem. Soc. 1979, 126, 2133.
20. Mériaudeau, P.; Ellestad, O.H.; Dufaux, M.; Naccache, C. J. Catal. 1982, 75, 243.
21. Chen, B.-H.; White, J.M. J. Phys. Chem. 1982, 86, 3534.
22. Brugniau, D.; Parker, S.D.; Read, G.E. Thin Solid Films 1984, 121, 247.

RECEIVED September 12, 1985

INDEXES

Author Index

Subject Index

A

B

Production and indexing by Keith B. Belton
Jacket design by Pamela Lewis

Elements typeset by Hot Type Ltd., Washington, DC
Printed and bound by Maple Press Co., York, PA